Journal of Applied Logics - IfCoLog Journal of Logics and their Applications

Volume 5, Number 9

December 2018

Disclaimer

Statements of fact and opinion in the articles in Journal of Applied Logics - IfCoLog Journal of Logics and their
Applications (JAL-FLAP) are those of the respective authors and contributors and not of the JAL-FLAP. Neither College Publications nor the JAL-FLAP make any representation, express or implied, in respect of the accuracy of the material in this journal and cannot accept any legal responsibility or liability for any errors or omissions that may be made. The reader should make his/her own evaluation as to the appropriateness or otherwise of any experimental technique described.

© Individual authors and College Publications 2018
All rights reserved.

ISBN 978-1-84890-294-7
ISSN (E) 2055-3714
ISSN (P) 2055-3706

College Publications
Scientific Director: Dov Gabbay
Managing Director: Jane Spurr

http://www.collegepublications.co.uk

All rights reserved. No part of this publication may be reproduced, stored in a retrieval system or transmitted in any form, or by any means, electronic, mechanical, photocopying, recording or otherwise without prior permission, in writing, from the publisher.

Editorial Board

Editors-in-Chief
Dov M. Gabbay and Jörg Siekmann

Marcello D'Agostino
Natasha Alechina
Sandra Alves
Arnon Avron
Jan Broersen
Martin Caminada
Balder ten Cate
Agata Ciabttoni
Robin Cooper
Luis Farinas del Cerro
Esther David
Didier Dubois
PM Dung
Amy Felty
David Fernandez Duque
Jan van Eijck

Melvin Fitting
Michael Gabbay
Murdoch Gabbay
Thomas F. Gordon
Wesley H. Holliday
Sara Kalvala
Shalom Lappin
Beishui Liao
David Makinson
George Metcalfe
Claudia Nalon
Valeria de Paiva
Jeff Paris
David Pearce
Brigitte Pientka
Elaine Pimentel

Henri Prade
David Pym
Ruy de Queiroz
Ram Ramanujam
Chrtian Retoré
Ulrike Sattler
Jörg Siekmann
Jane Spurr
Kaile Su
Leon van der Torre
Yde Venema
Rineke Verbrugge
Heinrich Wansing
Jef Wijsen
John Woods
Michael Wooldridge
Anna Zamansky

Scope and Submissions

This journal considers submission in all areas of pure and applied logic, including:

- pure logical systems
- proof theory
- constructive logic
- categorical logic
- modal and temporal logic
- model theory
- recursion theory
- type theory
- nominal theory
- nonclassical logics
- nonmonotonic logic
- numerical and uncertainty reasoning
- logic and AI
- foundations of logic programming
- belief revision
- systems of knowledge and belief
- logics and semantics of programming
- specification and verification
- agent theory
- databases
- dynamic logic
- quantum logic
- algebraic logic
- logic and cognition
- probabilistic logic
- logic and networks
- neuro-logical systems
- complexity
- argumentation theory
- logic and computation
- logic and language
- logic engineering
- knowledge-based systems
- automated reasoning
- knowledge representation
- logic in hardware and VLSI
- natural language
- concurrent computation
- planning

This journal will also consider papers on the application of logic in other subject areas: philosophy, cognitive science, physics etc. provided they have some formal content.

Submissions should be sent to Jane Spurr (jane.spurr@kcl.ac.uk) as a pdf file, preferably compiled in LaTeX using the IFCoLog class file.

CONTENTS

ARTICLES

Preface to the Special Issue of the 48th IEEE International Symposium on Multiple Valued Logic 1777
Robert Wille and Martin Lukac

Editor's Note 1779
Martin Lukac

Higher-Radix Chrestenson Gates for Photonic Quantum Computation . . 1781
Kaitlin N. Smith, Tim P. LaFave Jr., Duncan L. MacFarlane and Mitchell A. Thornton

Amoeba-inspired Electronic Computing System and its Application to Autonomous Walking of a Multi-legged Robot 1799
Kenta Saito, Naoki Suefuji, Seiya Kasai and Masashi Aono

Characterizing Parallel Multipliers for Detecting Hardware Trojans 1815
Akira Ito, Rei Ueno, Naofumi Homma and Takafumi Aoki

A Noise-shaping Analog-to-Digital Converter using a $\Delta\Sigma$ Modulator Feedforward Network 1833
Takao Waho

An Exact Optimization Method using ZDDs for Linear Decomposition
of Symmetric Index Generation Functions 1849
Shinobu Nagayama, Tsutomu Sasao and Jon Butler

Preface to the Special Issue of the 48th IEEE International Symposium on Multiple Valued Logic

Robert Wille and Martin Lukac

It is our great pleasure to present you the Selected Papers from the 48th IEEE International Symposium on Multiple-Valued Logic (ISMVL) held in Linz, Austria on May 16th-18th, 2018.

Multiple-valued logic still finds significant interest in the scientific community and, for almost 50 years, ISMVL is the main platform to present and discuss new trends as well as recent work in this domain. In the past decades, researchers in this domain as well as bordering areas have always been very active and we are glad that this did not change in the last year. In fact, with a total of 80 submissions, we received an impressive amount of papers covering recent findings and developments within this area.

This special issue is devoted to this symposium and contains the papers that have been evaluated best by both, the members of the program committee as well as the participants of the symposium.

The papers presented here reflect the wide range of topics which have been covered at ISMVL 2014 including such as algebra, security, logic design, hardware design, quantum computation, and more. This portfolio provides an in-depth view on the current developments in our domain which surely will have a significant impact on future development.

We would like to thank the members of the program committee and the external reviewers for their hard work in evaluating the submissions and providing detailed feedback and further suggestions to the authors.

Robert Wille
General Chair

Martin Lukac
PC Chair

October 31, 2018

Editor's Note

Martin Lukac

Multiple-Valued Logic (MVL) is a very broad area that encompasses theory, tools and realization of logic functions and its associated technology. As such MVL can be seen as starting point or many now existing areas such as higher Radix Logic, Fuzzy Logic, Quantum Logic or even Reversible Logic. Because of this very large field of influence, the selected papers in this special issue from the 48th International Symposium on Multiple-valued Logic also cover a very large and various areas of application of Multiple-Valued Logic.

The selected papers covers topics such as biologically inspired computing, quantum computing, function representation security and AD conversion. The papers are ordered in order of increases theoretical content but in general most of the papers are on the edge of the MVL research area.

The first paper studies the usage of a optical beam splitter with application to quantum computing. The main point of this paper is that it uses an optical component, a beam splitter, that was originally intended for optical communication and extends its usage to quantum optical computing approach. The components is used to simulate the quaternary extension of the Chrestenson quantum gate and the provided analysis shows the applicability of the evaluated component.

The second paper, deals with an biologically inspired hardware approach to solving the SAT problem as well as autonomous walking of an quadruped. The paper is interesting in the formulation of the amoeba movement into an constraint-satisfaction problem solver. Using a set of rules it is used to both solve SAT problem as well as a dynamically changing the problem of autonomous walking. While this paper does not use multiple-valued logic, it is a very good example of the adaptation of biological rules to the digital problem solving.

The third paper deals with hardware security and describe an analytic approach to determine the best approach in detecting hardware path-delay Trojan horses. The approach analyzes the least critical path; such path in the circuit that is the least likely used in the logic circuit. In this paper the analyzed circuits are strictly binary but the analysis is in the time domain and as such it is considered as not strictly binary problem solution.

The fourth paper is an application of the artificial neural network paradigms in the design of efficient AD converters. The AD converters are a very important part of any device that requires the conversion from analog to digital signal (or vice versa). This particular work describes an approach to increase the accuracy of AD conversion by using a Neural Network architecture allowing to increase the effective number of bits for relatively small circuit overhead.

Finally, the last paper deals with the minimization of the indexing logic functions, but analyzing symmetric indexing functions and providing an efficient design methodology. The proposed method uses ZDDs (Zero Suppressed Decision Diagrams) to represent the space of searched indexing functions which provides a considerable speedup and reduces the complexity of the search.

Martin Lukac
Guest Editor

October 31, 2018

Higher-Radix Chrestenson Gates for Photonic Quantum Computation

Kaitlin N. Smith, Tim P. LaFave Jr., Duncan L. MacFarlane, and Mitchell A. Thornton
Quantum Informatics Research Group
Southern Methodist University
Dallas, TX, USA
{knsmith, tlafave, dmacfarlane, mitch}@smu.edu

Abstract

A recently developed four-port coupler used in optical signal processing applications is shown to be equivalent to a Chrestenson operator, or gate, in radix-4 quantum information processing (QIP) applications. The radix-4 qudit is implemented as a location-encoded photon incident on one of the four ports of the coupler. The quantum informatics transfer matrix is derived for the device based upon the conservation of energy equations when the coupler is employed in a classical sense in an optical communications environment. The resulting transfer matrix is the radix-4 Chrestenson transform, and this operator is capable of placing a radix-4 qudit in a state of maximal superposition. This result indicates that a new practical device is available for use in the implementation of radix-4 QIP applications or in the construction of a radix-4 quantum computer.

1 Introduction

In the field of quantum computing, theory has been well developed over the last half-century. There is still room for the discovery of new operators and algorithms, but current work is enough to show the potential that quantum methods have for solving some of the most difficult scientific problems. Unfortunately, the physical implementations for QIP have not advanced as rapidly as the theory. Since a standard platform has not been chosen for the quantum computer (QC), efforts have been divided among many competing technologies with quantum optics being one of the more promising physical realizations. This work hopes to contribute a new, higher-radix component to the already well-established photonic quantum computing library.

The four-port coupler is an optical component shown theoretically to act as a quantum gate that can place a radix-4 photon-encoded qudit into a state of equal superposition. The gate realized optically is known as the radix-4 Chrestenson gate, and its transfer function is derived from the generalized radix-r Chrestenson transformation. Many QIP techniques require quantum superposition, so this operator is significant to the field of quantum computing due to the need to evolve information into a state of superposition for quantum algorithm execution. A description of the four-port coupler as well as a demonstration of its capabilities as a quantum optics operator will be shown mathematically in this work.

This paper proceeds as follows. A brief summary of important QIP concepts, details of the Chrestenson gate, with emphasis on the radix-4 implementation, and information about quantum optics are provided in Section 2. The four-port coupler, the component of interest, is described in Section 3. The physical realization of the four-port coupler with optical elements, including its fabrication and characterization, is included in Section 4. The demonstration of the four-port coupler as a radix-4 Chrestenson gate is presented in Section 5. Finally, a summary with conclusions can be found in Section 6. This paper is an extension of the work originally published in reference [1].

2 Quantum theory background

2.1 The qubit vs. qudit

The quantum bit, or qubit, is the standard unit of information for radix-2, or base-2, quantum computing. The qubit models information as a linear combination of two orthonormal basis states such as the states $|0\rangle$ and $|1\rangle$. $|0\rangle$ and $|1\rangle$ are Dirac notation representations where $|0\rangle = \begin{bmatrix} 1 & 0 \end{bmatrix}^T$ and $|1\rangle = \begin{bmatrix} 0 & 1 \end{bmatrix}^T$, respectively. The qubit differs from the classical bit by its ability to be in a state of superposition, or a state of linear combination, of all basis states. Superposition allows QIP algorithms to be very powerful since it allows for parallelism during computation so that multiple combinations of information can be evaluated at once. In other words, a single QC taking advantage of superposition can complete some tasks in a time frame that would require multiple classical computers working simultaneously. There are theoretically an infinite number of states for a qubit while in a state of superposition

$$|\Psi\rangle = x|0\rangle + y|1\rangle = \begin{bmatrix} x & y \end{bmatrix}^T \tag{1}$$

where x and y are complex values, $c \in \mathbb{C}$, such that $c = a + ib$ where i is an imaginary number, $i^2 = -1$. For the qubit $|\Psi\rangle$, the probability that $|\Psi\rangle = |0\rangle$ is equal to $x^*x = |x|^2$ and the probability that $|\Psi\rangle = |1\rangle$ is equal to $y^*y = |y|^2$ where the symbol * indicates a complex conjugate. The total probability of occupying either one basis state or the other must total to 100%, so the inner product, or dot product, of $|\Psi\rangle$ with itself must equal 1. In other words, $x^*x + y^*y = 1$. Once a qubit is measured, it collapses into a basis state as defined by the eigenvectors of the measurement operator [2]. The measurement operation causes a qubit's state of superposition to be lost.

Qubits are the current standard for encoding data in QIP, but it is possible to have a quantum system of higher order. Increasing the radix during computation allows for higher density data to be transmitted because more information is stored in each fundamental unit of information, or digit, of the system [3]. A quantum unit of dimension, or radix, $r > 2$ is referred to as a qudit. In this paper, the radix-4 qudit using four orthonormal basis states is of interest. The set of basis states used for the radix-4 qudit includes the vectors $|0\rangle = \begin{bmatrix} 1 & 0 & 0 & 0 \end{bmatrix}^{\mathrm{T}}$, $|1\rangle = \begin{bmatrix} 0 & 1 & 0 & 0 \end{bmatrix}^{\mathrm{T}}$, $|2\rangle = \begin{bmatrix} 0 & 0 & 1 & 0 \end{bmatrix}^{\mathrm{T}}$, and $|3\rangle = \begin{bmatrix} 0 & 0 & 0 & 1 \end{bmatrix}^{\mathrm{T}}$. Just like the radix-2 qubit, the radix-4 qudit is not limited to having the value of only one of its four possible basis states. The qudit is capable of existing in a linear combination, or a state of superposition, of all four basis states, as demonstrated by

$$|\Phi\rangle = v|0\rangle + x|1\rangle + y|2\rangle + z|3\rangle = \begin{bmatrix} v & x & y & z \end{bmatrix}^{\mathrm{T}} \quad (2)$$

where v, x, y, and z are complex values. These coefficients can be multiplied by their respective complex conjugates in order to derive the probability that the radix-4 qudit is in a particular basis state. The basis state probabilities of $|\Phi\rangle$ must sum to 100%, so $v^*v + x^*x + y^*y + z^*z = 1$.

For a radix-4 quantum system to be physically realized, a methodology must exist for encoding four distinct qudit basis states. In reference [4], Rabi oscillations are utilized to create radix-r quantum systems. Orbital angular momentum (OAM) states of light could also be used to encode the qudit [5]. In this paper, a radix-4 quantum state will be created using the location of light as the information carrier. This technique builds on the concept of the quantum photonic dual-rail representation of the qubit in order to physically realize the radix-4 qudit with a quad-rail implementation.

2.2 Quantum operations

According to the most popular quantum computing paradigm proposed in reference [6], a quantum state must be prepared by a QC in a known basis. Afterwards, meaningful information is generated by evolving to the quantum state through quantum operations. After all computations are complete, the quantum state must be able to be measured to produce an output. If a quantum algorithm is modeled as a circuit, quantum operations can be viewed as quantum logic gates. Each of these gates is represented by a unique, unitary transfer function matrix, $\mathbf{U}$, that is characterized by the following properties:

- $\mathbf{U}^\dagger \mathbf{U} = \mathbf{U}\mathbf{U}^\dagger = \mathbf{I}_r$
- $\mathbf{U}^{-1} = \mathbf{U}^\dagger$
- $\text{Rank}(\mathbf{U}) = r$
- $|\mathbf{U}| = 1$

When considering radix-r quantum operations, or gates, the transfer function matrices will always be square matrices each of a dimension that is a power of r. Therefore, radix-4 qudit operations will have a dimension that is a power of four, 4^k, where the power, k, indicates the amount of qudits transformed by the quantum operation.

2.3 The Chrestenson gate

The power of QIP lies in the ability for a quantum state to be in superposition, and achieving states of maximal superposition is of especially high importance because it is typically one of the first steps required in quantum algorithms. When the probability amplitudes are all nonzero and the square of their magnitudes are equivalent, the qubit or qudit is said to be maximally superimposed, or is in maximal superposition, with respect to some basis set. Practically, this means that the qubit or qudit is equally likely to be measured at the value of any of the possible basis vectors.

In radix-2 quantum computation, the Hadamard gate

$$\mathbf{H} = \frac{1}{\sqrt{2}} \begin{bmatrix} 1 & 1 \\ 1 & -1 \end{bmatrix} \tag{3}$$

is an important operator used to put a qubit in a maximally superimposed state. When a qubit originally in a basis state passes through the Hadamard gate, the transformed quantum information has equal probability of being observed, or measured, as either $|0\rangle$ or $|1\rangle$. Quantum operators exist for many different computation bases, such as radix-3 and above, that achieve equal, and therefore maximal, superposition among the corresponding basis states. These operators are derived using the discrete Fourier transform on Abelian groups. General theory of Fourier transforms on Abelian groups is outlined in the literature [7, 8]. The multiple-valued generalization of the radix-2 quantum Hadamard gate and its transfer matrix is composed of discretized versions of the orthogonal Chrestenson basis function set [8]. This QIP gate is generally referred to as the Chrestenson gate [9]. Examples of useful applications of the Chrestenson transform in QIP can be found in reference [10].

The Chrestenson operator, as the generalized version of the Hadamard operator, has a shape that depends on the radix of computation. The resulting radix-r Chrestenson transformation matrix for a single qudit has a size of $r \times r$, and the basis vectors that span the matrix are composed of r different values. Since the Chrestenson transformation matrix is normalized with a scalar factor, $\frac{1}{\sqrt{r}}$, and the matrix is orthogonal, both the column and row vectors of the operator form an orthonormal set. Each of the components within a Chrestenson transform matrix is one of the rth roots of unity raised to an integral power [8, 9]. The rth roots of unity can be visualized as r points that are evenly-spaced on the unit circle in the complex plane. The roots of unity are indicated as w_k where $k = 0, 1, ..., (r-1)$, and the point (1,0), denoted as w_0, is always included in this set. Each root satisfies $(w_k)^r$ as roots of one. The closed-form representation of the rth roots of unity for a radix-r Chrestenson transformation is

$$w_k = e^{i\frac{2\pi}{r} \times k}. \qquad (4)$$

Fig. 1 contains the plots for the rth roots of unity for $r = 2, 3, 4,$ and 5.

The structure of the Chrestenson transform matrix is in the form of a Vandermonde matrix where each row vector consists of a rth root of unity, w_k, raised to an integral power j. Each element of the matrix is some form of w_k^j where j is the column index and k is the row index. In this indexing scheme, $j = 0$ is assigned to the leftmost column vector and $j = (r-1)$ is assigned to the rightmost column vector. Similarly, $k = 0$ is assigned to the topmost row vector and $k = (r-1)$ is assigned to the bottommost row vector. It can be observed that the Hadamard matrix results from the Chrestenson transform matrix with $r = 2$, confirming that the Chrestenson transform is a generalization of the Hadamard transform for higher-dimensional quantum systems. The generalized radix-r Chrestenson transform matrix, $\mathbf{C_r}$, is

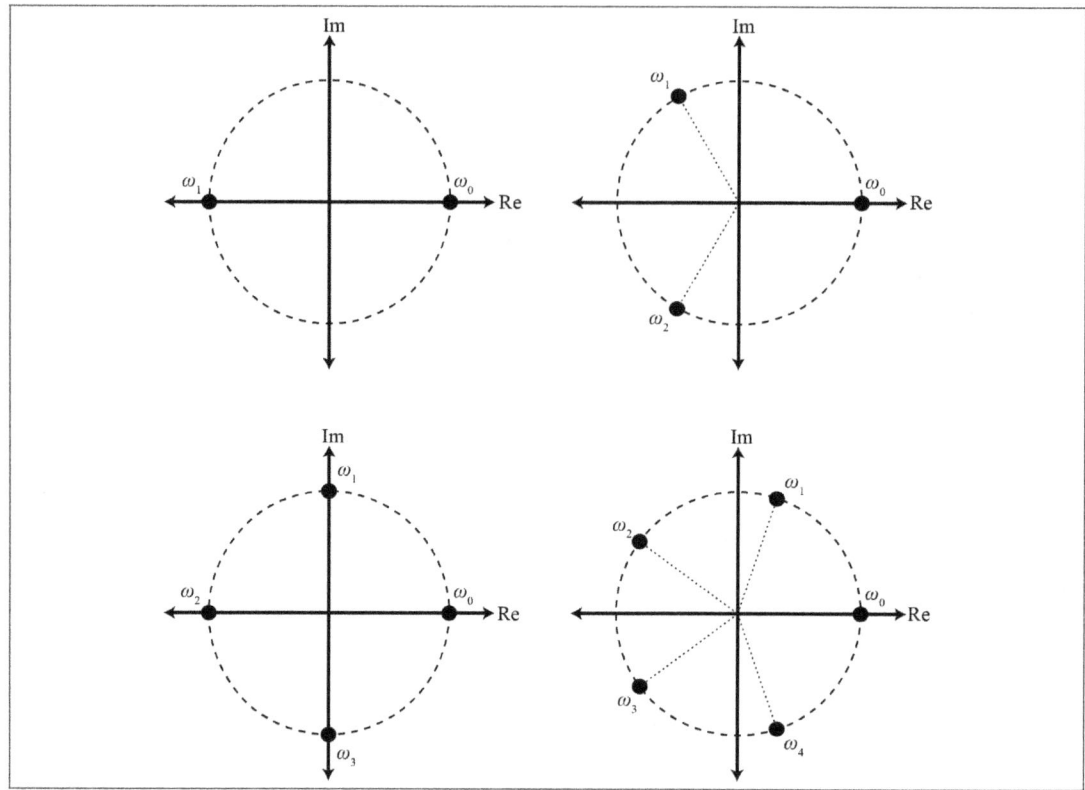

Figure 1: Roots of unity in the complex plane for $r = 2, 3, 4$, and 5.

represented with a matrix in the form of

$$\mathbf{C}_r = \frac{1}{\sqrt{r}} \begin{bmatrix} w_0^0 & w_0^1 & \cdots & w_0^{(r-1)} \\ w_1^0 & w_1^1 & \cdots & w_1^{(r-1)} \\ \vdots & \vdots & \ddots & \vdots \\ w_{(r-1)}^0 & w_{(r-1)}^1 & \cdots & w_{(r-1)}^{(r-1)} \end{bmatrix}. \qquad (5)$$

Using the fourth roots of unity, $w_0 = \exp[(i2\pi/4)*0] = 1$, $w_1 = \exp[(i2\pi/4)*1] = i$, $w_2 = \exp[(i2\pi/4)*2] = -1$, and $w_3 = \exp[(i2\pi/4)*3] = -i$, in Eq. 5, the radix-4

Chrestenson gate transfer matrix becomes

$$\mathbf{C_4} = \frac{1}{\sqrt{4}} \begin{bmatrix} 1 & 1 & 1 & 1 \\ 1 & i & -1 & -i \\ 1 & -1 & 1 & -1 \\ 1 & -i & -1 & i \end{bmatrix}. \tag{6}$$

The radix-4 Chrestenson gate ($\mathbf{C_4}$), allows a radix-4 qudit originally in a basis to evolve into a quantum state of equal superposition. The following example shows how the radix-4 qudit $|a\rangle = |0\rangle$ evolves to $|b\rangle = \frac{1}{2}|0\rangle + \frac{1}{2}|1\rangle + \frac{1}{2}|2\rangle + \frac{1}{2}|3\rangle$, taking the value of the first column of the radix-4 Chrestenson matrix, after passing through the $\mathbf{C_4}$ transform

$$\mathbf{C_4}|a\rangle = |b\rangle,$$

$$\mathbf{C_4}|0\rangle = \frac{1}{\sqrt{4}} \begin{bmatrix} 1 & 1 & 1 & 1 \\ 1 & i & -1 & -i \\ 1 & -1 & 1 & -1 \\ 1 & -i & -1 & i \end{bmatrix} \begin{bmatrix} 1 \\ 0 \\ 0 \\ 0 \end{bmatrix} = \frac{1}{2} \begin{bmatrix} 1 \\ 1 \\ 1 \\ 1 \end{bmatrix},$$

$$\mathbf{C_4}|0\rangle = \frac{1}{2}[|0\rangle + |1\rangle + |2\rangle + |3\rangle].$$

If $|a\rangle = |3\rangle$, the radix-4 qudit would evolve to $|b\rangle = \frac{1}{2}|0\rangle - \frac{1}{2}i|1\rangle - \frac{1}{2}|2\rangle + \frac{1}{2}i|3\rangle$, taking the value of the last column of the $\mathbf{C_4}$ transformation matrix.

$$\mathbf{C_4}|3\rangle = \frac{1}{\sqrt{4}} \begin{bmatrix} 1 & 1 & 1 & 1 \\ 1 & i & -1 & -i \\ 1 & -1 & 1 & -1 \\ 1 & -i & -1 & i \end{bmatrix} \begin{bmatrix} 0 \\ 0 \\ 0 \\ 1 \end{bmatrix} = \frac{1}{2} \begin{bmatrix} 1 \\ -i \\ -1 \\ i \end{bmatrix},$$

$$\mathbf{C_4}|3\rangle = \frac{1}{2}[|0\rangle - i|1\rangle - |2\rangle + i|3\rangle].$$

The schematic symbol of the $\mathbf{C_4}$ gate is pictured in Fig. 2. This symbol can be used in radix-4 quantum circuit diagrams.

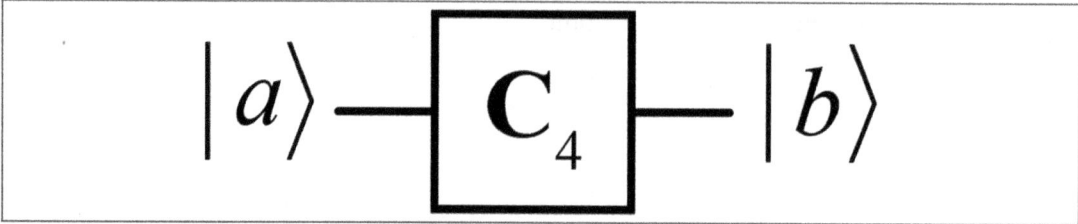

Figure 2: Symbol of the radix-4 Chrestenson gate, C_4.

2.4 Quantum optics

Optical quantum implementations are among the more successful physical realizations of quantum states. In these systems, orthogonal basis states can be encoded into photon OAM states, polarization, or location, and the state can easily evolve by passing through linear optical elements. The photon resists coupling to other objects in its environment, allowing it to maintain its quantum state and not decohere for long periods of time [11]. Additionally, the ability to maintain coherence enables the photon to travel great distances at room temperature, making it a good candidate for long-haul quantum information transmission.

Although photons offer the benefit of state stability in QIP applications, their failure to interact with their surroundings prevents them from coupling with each other. Photon-to-photon interaction is difficult, limiting the development of reliable controlled multi-qubit, or multi-qudit in higher radix systems, gate implementations. Without operations such as the radix-2 controlled-NOT (CNOT) gate or the controlled-phase gate, a functional QC cannot exist.

It was once thought that photonic quantum computation was unachievable without nonlinear optical elements, but the presentation of the KLM protocol in reference [12] improved the outlook for quantum optics. In that work, a methodology for implementing photonic multi-qubit operations using linear optics was introduced. These multi-qubit photonic gates, however, are unfortunately limited by probabilistic operation. Currently, the two-qubit CNOT operation can only work 1/4 of the time when implemented with linear optical elements in the best case scenario [13].

The subject of this paper is a photonic radix-4 Chrestenson gate. Since this quantum operator is formed from linear optical elements and transforms a single qudit at a time, the gate is theoretically deterministic in nature.

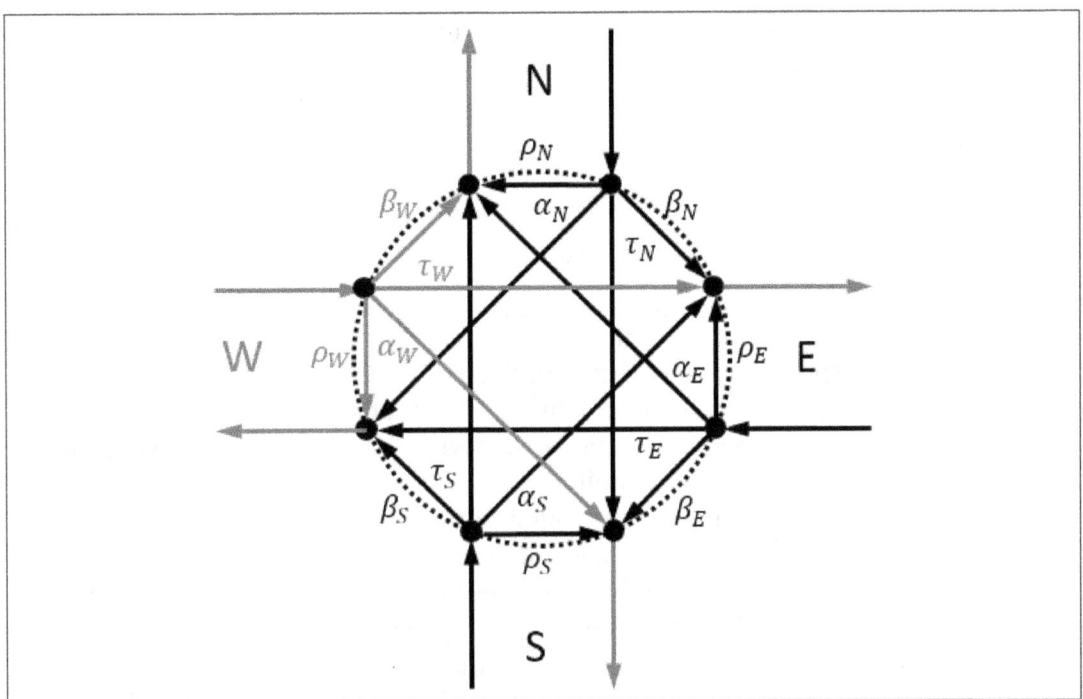

Figure 3: Signal flow for four-port coupler with input at W.

3 The four-port coupler

The four-port coupler is an optical component introduced and described in reference [14]. This device is composed of four inputs and four outputs where the input and output are referred to by their orientation on the component of either W, N, E, or S. When a single beam is sent into one of the coupler inputs, the component routes a fraction of the original signal to each of the four outputs. This beam division is caused by the transmission and reflection of signals within the coupler. Each fraction of the input beam seen at an output corresponds to one of the following components of the original signal: a reflected component ρ, a transmitted component τ, a right-directed component α, and a left-directed component β. An illustration of signal flow of the four-port coupler can be seen in Fig. 3. This image is recreated from a figure included in reference [14].

Fig. 3 demonstrates in blue a signal entering the four-port coupler from the W port and exiting the component from the W, N, E, and S ports. The output signals are generated by ρ_W, β_W, τ_W, and α_W, respectively. Whenever a single input enters the component, all four coupling coefficients are generated to produce four outputs.

The coupling coefficients produced with a particular port input can be derived with the coupling coefficient matrix,

$$\begin{bmatrix} \rho_W & \alpha_N & \tau_E & \beta_S \\ \beta_W & \rho_N & \alpha_E & \tau_S \\ \tau_W & \beta_N & \rho_E & \alpha_S \\ \alpha_W & \tau_N & \beta_E & \rho_S \end{bmatrix}. \tag{7}$$

To produce the outputs, an input vector taking the form of $[WNES]^\mathrm{T}$ is multiplied by the matrix in Eq. 7 to create a column vector of coupling coefficients. The produced output column vector also takes the form of $[WNES]^\mathrm{T}$. The composition of the output vector in terms of coupling coefficients indicates what portion of the input signal contributes to an output from a port.

The four-port coupler does not consume nor dissipate any of the energy that is input into the component. Therefore, to conserve energy, all of the energy entering the element must be equal to the energy leaving the element. This concept leads to the creation of equations that act as conditions that must hold true for energy conservation. The 10 energy conservation equations of Eqs. 8-17, first derived in reference [14], use the coupling coefficients found in the matrix of Eq. 7. These equations are:

$$\rho_W^* \rho_W + \beta_W^* \beta_W + \tau_W^* \tau_W + \alpha_W^* \alpha_W = 1, \tag{8}$$

$$\rho_N^* \rho_N + \beta_N^* \beta_N + \tau_N^* \tau_N + \alpha_N^* \alpha_N = 1, \tag{9}$$

$$\rho_E^* \rho_E + \beta_E^* \beta_E + \tau_E^* \tau_E + \alpha_E^* \alpha_E = 1, \tag{10}$$

$$\rho_S^* \rho_S + \beta_S^* \beta_S + \tau_S^* \tau_S + \alpha_S^* \alpha_S = 1, \tag{11}$$

$$\rho_W^* \tau_E + \beta_W^* \alpha_E + \tau_W^* \rho_E + \alpha_W^* \beta_E = 0, \tag{12}$$

$$\alpha_N^* \beta_S + \rho_N^* \tau_S + \beta_N^* \alpha_S + \tau_N^* \rho_S = 0, \tag{13}$$

$$\rho_W^* \alpha_N + \beta_W^* \rho_N + \tau_W^* \beta_N + \alpha_W^* \tau_N = 0, \tag{14}$$

$$\alpha_N^* \tau_E + \rho_N^* \alpha_E + \beta_N^* \rho_E + \tau_N^* \beta_E = 0, \qquad (15)$$

$$\tau_E^* \beta_S + \alpha_E^* \tau_S + \rho_E^* \alpha_S + \beta_E^* \rho_S = 0, \qquad (16)$$

and

$$\rho_W^* \beta_S + \beta_W^* \tau_S + \tau_W^* \alpha_S + \alpha_W^* \rho_S = 0. \qquad (17)$$

The first four conditions seen in Eqs. 8-11 exist since the inner product of each produced field vector from a single input, W, N, E, and S, with itself must sum to 1 for energy conservation. The last six conditions seen in Eqs. 12-17 exist due to energy conservation that occurs whenever two inputs are present in the component. Since the coefficient vectors are orthogonal, the inner product between the two produced coupling coefficient vectors corresponding to inputs at two different ports must equal zero. There are only 6 constraints produced from sending two inputs to the four-port coupler because the input combinations are commutative (i.e. $AB = BA$). The cases of three inputs and four inputs into the four-port coupler do not create additional constraints, so they are omitted [14].

4 Physical realizations of the four-port coupler

A macroscopic realization of a four-port coupler is shown in Fig. 4. Whereas a popular implementation of a radix-2 Hadamard gate is an optical beam splitter, polarizing or not, the macroscopic four-port coupler is a unitary extension of a two-prism beam splitting cube. Here, the macroscopic four-port coupler is comprised of four right angle prisms, coated with an appropriate thin film, cemented together with care given to the precise mating of the four prism corners. This component has been used to demonstrate novel, four leg Michelson interferometers designed in reference [15].

Integrated photonic four-port couplers were previously demonstrated for applications in optical signal processing as part of a two-dimensional array of waveguides in a multi-quantum well (MQW) GaInAsP indium phosphide (InP) architecture [16, 17]. Fig. 5 shows an electron micrograph of a coupler fabricated at the intersection of two ridge waveguides.

The optical behavior of the four-port coupler depends on frustrated total internal reflection [18]. The evanescent field of light incident on the coupler is transfered across the width of the coupler that may be an air gap or a thin slice of dielectric. Provided the barrier width is small enough, a part of the exponentially decaying

Figure 4: Macroscopic realization of a four-port coupler.

optical power of the incident light is transmitted across while the remaining optical power is reflected. Thus, a fraction of light incident on a four-port coupler may be transmitted to the ongoing waveguide, reflected to both perpendicular waveguides, or reflected back into the originating waveguide. The fractions of light in outbound waveguides are determined by the refractive indices of the waveguide and coupler materials in addition to the width of the coupler.

4.1 Fabrication

Fabrication of the coupler was performed in several steps using nanoelectronic processing techniques. First, coupler regions of 180 nm widths and 7 μm lengths were defined by patterning a thin metallic chromium mask layer atop the waveguides by focused ion beam (FIB) lithography. Precision alignment and orientation of the coupler to the waveguides during FIB processing was achieved with alignment markers fabricated beforehand with the waveguides using conventional microelectronic processing steps. High aspect ratio trenches were then etched using a hydrogen bromine (HBr) based [15] inductively coupled plasma (ICP) to a depth of 3.9 μm. This depth

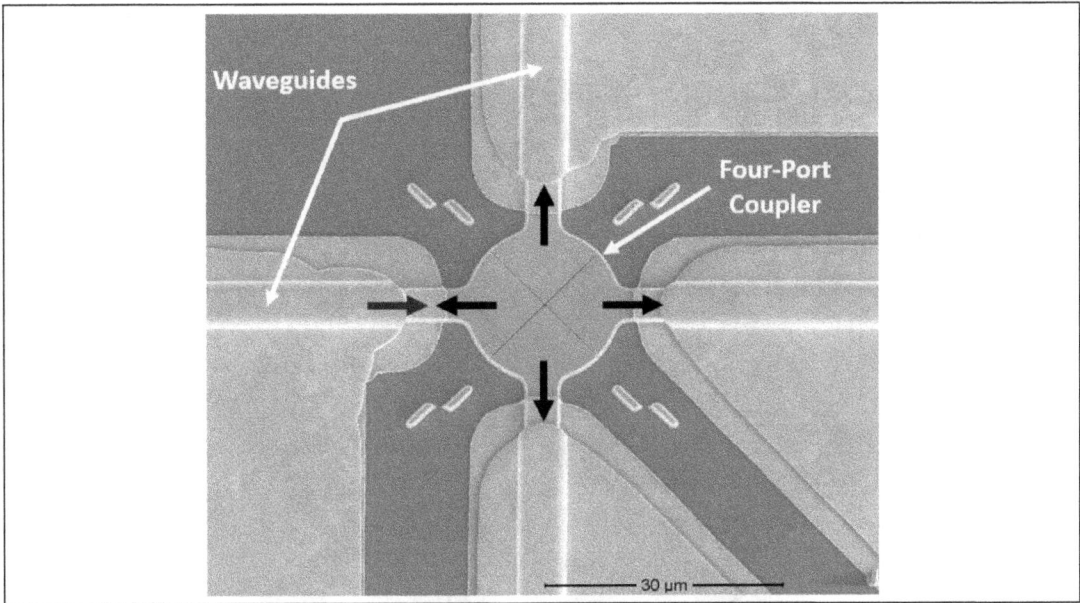

Figure 5: Cross sectional scanning electron microscope image of a four-port coupler in MQW-InP.

allows the coupler to fully cover optical modes confined to the quantum well region of the waveguides.

The optimal air gap width for 25% power on all output waveguides of about 90 nm was slightly smaller than the processing capability of the ICP dry etch tool for the required high-aspect ratio etch. Consequently, to meet this requirement for a wavelength of 1550 nm, alumina (Al_2O_3), with a refractive index of $n = 1.71$, was back-filled into the trench using atomic layer deposition (ALD). The resulting alumina-filled trench is shown in the composite cross-sectional transmission electron micrographs in Fig. 6.

4.2 Characterization

A 1550 nm laser was coupled into the waveguides using a tapered lens fiber at one input port. The near-field modes of light were coupled out of the device and into another tapered lens fiber for optical power measurement to characterize the coupling efficiency of the four-port coupler. The measured average power coefficients were $\alpha = 0.156$, $\beta = 0.140$, $\rho = 0.302$, and $\tau = 0.220$ for a measured total average coupling efficiency of 82% for the four-port coupler [16].

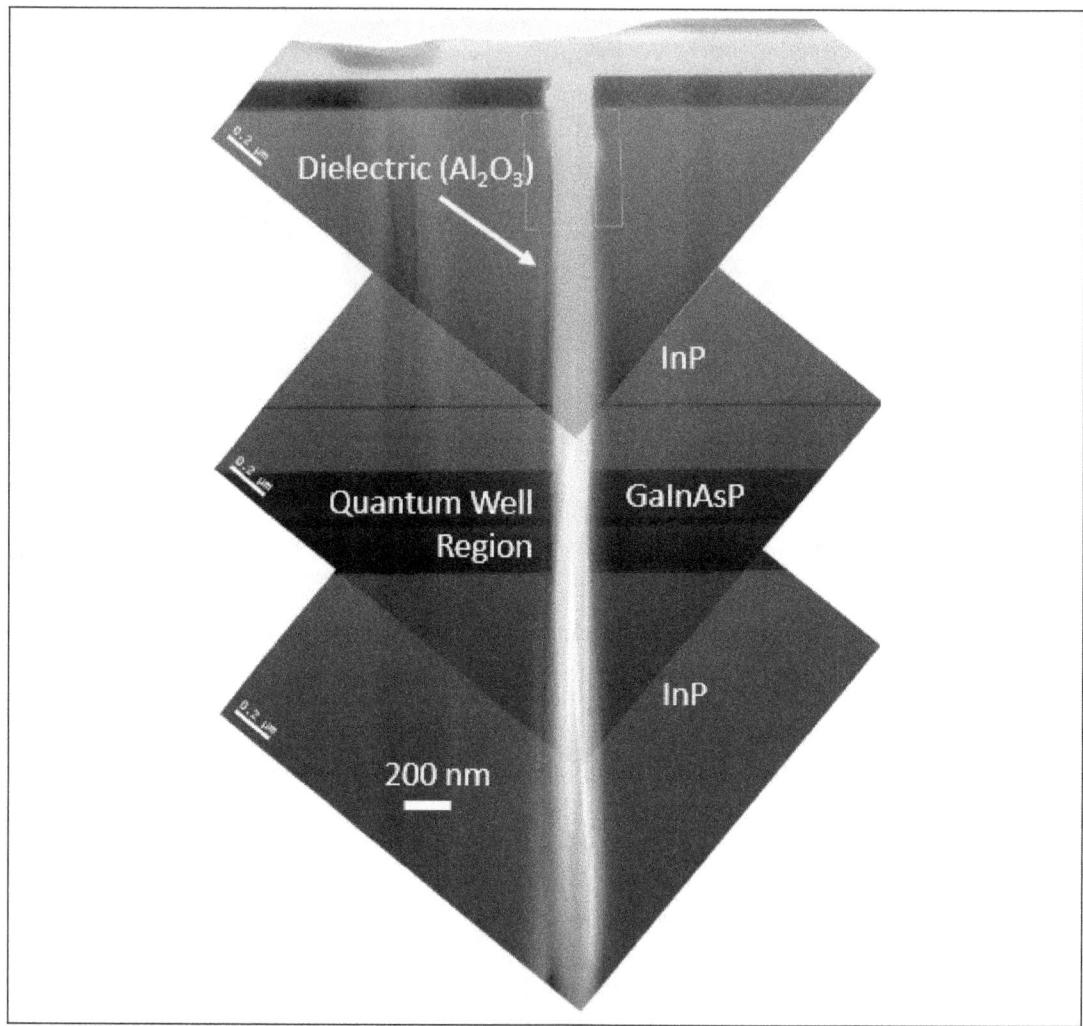

Figure 6: Cross sectional transmission electron micrograph of a four-port coupler backfilled with alumina using atomic layer deposition.

5 Implementing qudit quantum operations with the coupler

It is known that the Hadamard gate meant for use with a quantum qubit can be constructed from a beam splitter [11]. The radix-4 Chrestenson operation, an operation that acts on a quantum encoding using four basis states, transforms a radix-4 qudit, and the four-port coupler is a physical realization of this gate. In the realization of the radix-4 Chrestenson gate, the ports of the four-port coupler must be encoded in order to represent the four qudit basis states. In this paper, the following encoding has been chosen for the location-based scheme: port W is the $|0\rangle$ rail, port N is the $|1\rangle$ rail, port E is the $|2\rangle$ rail, and port S is the $|3\rangle$ rail.

The four-port coupler follows 10 energy conservation equations, Eqs. 8-17, that are algebraically nonlinear. If the radix-4 Chrestenson matrix values are substituted for the values of the coupling coefficients in Eq. 7, the energy conservation constraints are satisfied and the following matrix is generated:

$$\begin{bmatrix} \rho_W = \frac{1}{2} & \alpha_N = \frac{1}{2} & \tau_E = \frac{1}{2} & \beta_S = \frac{1}{2} \\ \beta_W = \frac{1}{2} & \rho_N = \frac{1}{2}i & \alpha_E = -\frac{1}{2} & \tau_S = -\frac{1}{2}i \\ \tau_W = \frac{1}{2} & \beta_N = -\frac{1}{2} & \rho_E = \frac{1}{2} & \alpha_S = -\frac{1}{2} \\ \alpha_W = \frac{1}{2} & \tau_N = -\frac{1}{2}i & \beta_E = -\frac{1}{2} & \rho_S = \frac{1}{2}i \end{bmatrix}.$$

When a single photon, representing a qudit, is applied to one of the inputs the four-port coupler, either W, N, E, or S, energy is conserved and the radix-4 Chrestenson transform is achieved. The photon leaves the gate with equal superposition of all basis states. In other words, the photon has a 1/4 probability of being located in any of the output ports W, N, E, or S representing the basis states $|0\rangle$, $|1\rangle$, $|2\rangle$, or $|3\rangle$, respectively:

$$\rho_W^* \rho_W + \beta_W^* \beta_W + \tau_W^* \tau_W + \alpha_W^* \alpha_W = 1,$$

$$\left(\frac{1}{2}\right)\left(\frac{1}{2}\right) + \left(\frac{1}{2}\right)\left(\frac{1}{2}\right) + \left(\frac{1}{2}\right)\left(\frac{1}{2}\right) + \left(\frac{1}{2}\right)\left(\frac{1}{2}\right) = 1,$$

$$\rho_N^* \rho_N + \beta_N^* \beta_N + \tau_N^* \tau_N + \alpha_N^* \alpha_N = 1,$$

$$\left(-\frac{1}{2}i\right)\left(\frac{1}{2}i\right) + \left(-\frac{1}{2}\right)\left(-\frac{1}{2}\right) + \left(\frac{1}{2}i\right)\left(-\frac{1}{2}i\right) + \left(\frac{1}{2}\right)\left(\frac{1}{2}\right) = 1,$$

$$\rho_E^* \rho_E + \beta_E^* \beta_E + \tau_E^* \tau_E + \alpha_E^* \alpha_E = 1,$$

$$\left(\frac{1}{2}\right)\left(\frac{1}{2}\right) + \left(-\frac{1}{2}\right)\left(-\frac{1}{2}\right) + \left(\frac{1}{2}\right)\left(\frac{1}{2}\right) + \left(-\frac{1}{2}\right)\left(-\frac{1}{2}\right) = 1,$$

$$\rho_S^* \rho_S + \beta_S^* \beta_S + \tau_S^* \tau_S + \alpha_S^* \alpha_S = 1,$$

$$\left(-\frac{1}{2}i\right)\left(\frac{1}{2}i\right) + \left(\frac{1}{2}\right)\left(\frac{1}{2}\right) + \left(\frac{1}{2}i\right)\left(-\frac{1}{2}i\right) + \left(-\frac{1}{2}\right)\left(-\frac{1}{2}\right) = 1.$$

If two signals are input into the four-port coupler Chrestenson gate, the conservation of energy causes the inner product of the two produced vectors of coupling coefficients to be zero:

$$\rho_W^* \tau_E + \beta_W^* \alpha_E + \tau_W^* \rho_E + \alpha_W^* \beta_E = 0,$$

$$\left(\frac{1}{2}\right)\left(\frac{1}{2}\right) + \left(\frac{1}{2}\right)\left(-\frac{1}{2}\right) + \left(\frac{1}{2}\right)\left(\frac{1}{2}\right) + \left(\frac{1}{2}\right)\left(-\frac{1}{2}\right) = 0,$$

$$\alpha_N^* \beta_S + \rho_N^* \tau_S + \beta_N^* \alpha_S + \tau_N^* \rho_S = 0,$$

$$\left(\frac{1}{2}\right)\left(\frac{1}{2}\right) + \left(-\frac{1}{2}i\right)\left(-\frac{1}{2}i\right) + \left(-\frac{1}{2}\right)\left(-\frac{1}{2}\right) + \left(\frac{1}{2}i\right)\left(\frac{1}{2}i\right) = 0,$$

$$\rho_W^* \alpha_N + \beta_W^* \rho_N + \tau_W^* \beta_N + \alpha_W^* \tau_N = 0,$$

$$\left(\frac{1}{2}\right)\left(\frac{1}{2}\right) + \left(\frac{1}{2}\right)\left(\frac{1}{2}i\right) + \left(\frac{1}{2}\right)\left(-\frac{1}{2}\right) + \left(\frac{1}{2}\right)\left(-\frac{1}{2}i\right),$$

$$\alpha_N^* \tau_E + \rho_N^* \alpha_E + \beta_N^* \rho_E + \tau_N^* \beta_E = 0,$$

$$\left(\frac{1}{2}\right)\left(\frac{1}{2}\right) + \left(-\frac{1}{2}i\right)\left(-\frac{1}{2}\right) + \left(-\frac{1}{2}\right)\left(\frac{1}{2}\right) + \left(\frac{1}{2}i\right)\left(-\frac{1}{2}\right) = 0,$$

$$\tau_E^* \beta_S + \alpha_E^* \tau_S + \rho_E^* \alpha_S + \beta_E^* \rho_S = 0,$$

$$\left(\frac{1}{2}\right)\left(\frac{1}{2}\right) + \left(-\frac{1}{2}\right)\left(-\frac{1}{2}i\right) + \left(\frac{1}{2}\right)\left(-\frac{1}{2}\right) + \left(-\frac{1}{2}\right)\left(\frac{1}{2}i\right) = 0,$$

$$\rho_W^* \beta_S + \beta_W^* \tau_S + \tau_W^* \alpha_S + \alpha_W^* \rho_S = 0,$$

$$\left(\frac{1}{2}\right)\left(\frac{1}{2}\right) + \left(\frac{1}{2}\right)\left(-\frac{1}{2}i\right) + \left(\frac{1}{2}\right)\left(-\frac{1}{2}\right) + \left(\frac{1}{2}\right)\left(\frac{1}{2}i\right) = 0.$$

Since these equations are satisfied with the elements of the derived radix-4 Chrestenson transform matrix, the four-port coupler proves to act as an effective radix-4 Chrestenson gate.

6 Conclusion

In this paper, an integrated photonic four-port coupler that has potential for integration in radix-4 qudit based quantum photonic circuits is discussed. By showing that the radix-4 Chrestenson transfer function satisfies the operational conditions imposed by the conservation of energy for the coupler, we demonstrate the component's ability to act as a radix-4 qudit Chrestenson gate in an optical quantum system. The Chrestenson gate puts a radix-r qudit into a state of equal superposition between all orthogonal basis states, so the discovery of a physical realization of such a gate is significant. Because a linear combination of radix-4 basis states can be achieved in quantum optics whenever the four-port coupler is used, QIP algorithms that utilize maximal qudit superposition can be realized with this element. The introduction of new quantum applications for the four-port coupler as a radix-4 Chrestenson gate will lead to additional gates and methods that make radix-4 quantum photonic systems more robust.

References

[1] K. N. Smith, T. P. LaFave, Jr., D. L. MacFarlane, and M. A. Thornton. A Radix-4 Chrestenson Gate for Optical Quantum Computation. *IEEE International Symposium on Multiple Valued Logic (ISMVL)*, pp. 260-265, 2018.

[2] M. A. Nielsen and I. L. Chuang. *Quantum Computation and Quantum Information.* Cambridge University Press, 2010.

[3] D. M. Miller and M. A. Thornton. *Multiple-Valued Logic Concepts and Representations.* Morgan & Claypool Publishers, 2008.

[4] K. Fujii. Quantum optical construction of generalized Pauli and Walsh-Hadamard matrices in three level systems. *arXiv preprint quant-ph/0309132*, 2003.

[5] J. C. García-Escartín and P. Chamorro-Posada. Quantum multiplexing with the orbital angular momentum of light. *Phys. Rev. A*, 78(6), 2008.

[6] D. Deutsch. Quantum theory, the Church-Turing principle and the universal quantum computer. *Proceedings of the Royal Society of London A: Mathematical, Physical and Engineering Sciences*, 400(1818): 97-117, 1985.

[7] N. Y. Vilenkin. Concerning a class of complete orthogonal systems. *Dokl. Akad. Nauk SSSR, Ser. Math*, 11, 1947.

[8] H. E. Chrestenson. A class of generalized Walsh functions. *Pacific Journal of Mathematics*, 5(1): 17-31, 1955.

[9] Z. Zilic and K. Radecka. Scaling and better approximating quantum Fourier transform by higher radices. *IEEE Trans. on Computers*, 56(2): 202-207, 2007.

[10] Z. Zilic and K. Radecka. The Role of Super-fast Transforms in Speeding up Quantum Computations. *IEEE International Symposium on Multiple Valued Logic (ISMVL)*, pp. 129-135, 2002.

[11] P. Kok, W. J. Munro, K. Nemoto, T. C. Ralph, J. P. Dowling, and G. J. Milburn. Linear optical quantum computing with photonic qubits. *Rev. of Mod. Phys.*, 79(1): 135-174, 2007.

[12] E. Knill, R. Laflamme, and G. J. Milburn. A scheme for efficient quantum computation with linear optics. *Nature*, 409(6816):46-52, 2001.

[13] J. Eisert. Optimizing linear optics quantum gates. *Phys. Rev. Lett.*, 95(4), 2005.

[14] D. L. MacFarlane, J. Tong, C. Fafadia, V. Govindan, L. R. Hunt, and I. Panahi. Extended lattice filters enabled by four-directional couplers. *Applied Optics*, 43(33): 6124-6133, 2004.

[15] N. Sultana, W. Zhou, T. LaFave Jr and D. L. MacFarlane. HBr based inductively coupled plasma etching of high aspect ratio nanoscale trenches in InP: Considerations for photonic applications. *J. Vac. Sci. B*, 27(6): 2351-2356, 2009.

[16] D. L. MacFarlane, M. P. Christensen, K. Liu, T. P. LaFave Jr., G. A. Evans, N. Sultana, T. W. Kim, J. Kim, J. B. Kirk, N. Huntoon, A. J. Stark, M. Dabkowski, L. R. Hunt, and V. Ramakrishna. Four-port nanophotonic frustrated total-internal reflection coupler. *IEEE Phot. Tech. Lett.*, 24(1):58-60, 2012.

[17] D. L. MacFarlane, M. P. Christensen, A. El Nagdi, G. A. Evans, L. R. Hunt, N. Huntoon, J. Kim, T. W. Kim, J. Kirk, T. P. LaFave Jr., K. Liu, V. Ramakrishna, M. Dabkowski, and N. Sultana. Experiment and theory of an active optical filter. *IEEE J. Quant. Electron.*, 48(3): 307-317, 2012.

[18] D. S. Gale. Frustrated total internal reflection. *Am. J. Phys.*, 40(7): 1038-1039, 1972.

Amoeba-inspired electronic computing system and its application to autonomous walking of a multi-legged robot

Kenta Saito, Naoki Suefuji, Seiya Kasai
Research Center for Integrated Quantum Electronics and Graduate School of Information Science & Technology, Hokkaido University
k-saito@rciqe.hokudai.ac.jp, kasai@rciqe.hokudai.ac.jp

Masashi Aono
Faculty of Environment and Information Studies, SFC, Keio University

Abstract

An amoeba-inspired electronic computing system that searches a solution of a combinational optimization problem was developed. The computing system, called electronic amoeba, electronically represents the spatiotemporal dynamics of a single-celled amoeboid organism that is trying to maximize its food intake while minimizing the risks. We implemented the system using a conventional electronic circuit and successfully demonstrated its solution search capability for a Boolean satisfiability problem, SAT. The electronic amoeba was applied to autonomous walking of a four-legged robot without programming any leg maneuvers. The robot could walk to the target direction by successively searching a combination of the multi-valued leg joint states depending on the previous states and sensor information. We also confirmed that our approach arose the ability to travel over obstacles without prior information.

1 Introduction

Autonomous control of robots and other machines needs intelligence that successively finds appropriate behaviors in various situations even when unknown events occur. Such intelligence is considered to arise from not only the machine learning

---

This work was partly supported by MEXT KAKENHI Grant Numbers JP25110001 and JP25110013, JSPS KAKENHI Grant Numbers JP16K14240 and JP18H01487.

but also the optimization; optimization makes it possible to find out an optimal action that satisfies the various demands and constraints. Considering the mobility of the robot, compact, fast, and low-power optimization computer is strongly required. The conventional von Neumann-type computer usually has to use long computation time and a lot of energy to solve the optimization problem: in general, there is no analytical method for solving the optimization problem and solution search technique is indispensable to find out a solution from the anomalous solution candidates. For efficient solution search, several physical systems in nature have been investigated recently. In particular, the systems based on Ising model [1–3] have been developed intensively, where the system computes with the help of the spin relaxation process in a ferromagnetic material. However, they have difficulties in compact implementation and problem mapping. Another potential approach for efficient solution search can be seen biology, because organisms in nature have learned to optimize their behaviors efficiently to survive in a harsh environment. An attractive example is *Physarum polycephalum*, a single-celled amoeboid organism. It possesses intelligence even though it has no a brain [4]. From the observation of the foraging behavior of the amoeboid organism, Aono *et al.* developed an amoeba-inspired solution search algorithm and confirmed its efficient solution capability [5, 6]. Based on the abstract model developed by them, we designed a simple analogue electronic circuit that represents its spatiotemporal dynamic behavior âĂIJelectronic amoebaâĂİ [7]. Recently, we successfully demonstrated its application to the autonomous walking of the multi-legged robot [8], whereas most of the autonomous robots have been controlled by the state machine and the machine learning [9]. In this paper, we describe the details of the system implementation and mapping the problem in the electronic amoeba, and discuss the relationship between the solution search time and the error intentionally imposed for searching. We also give explanation on the design of the constraints for searching leg maneuver in the multi-legged robot walking. Furthermore, we show that the optimization-based autonomous walking using the electronic amoeba arises an ability of traveling over unknown obstacles.

2 Concept and implementation

Figure 1 shows the basic concept of the electronic amoeba together with an amoeba-based computer as a counterpart. The amoeba-based computer shown in Fig. 1(a) utilizes an amoeba organism for solution search. A state variable x_i is the length of pseudopod that stretches and shrinks along the groove on the substrate. If the pseudopod i crosses the threshold, then $x_i = 1$, else $x_i = 0$. The electronic amoeba shown in Fig. 1(b) consists of an analogue electronic circuit representing the spa-

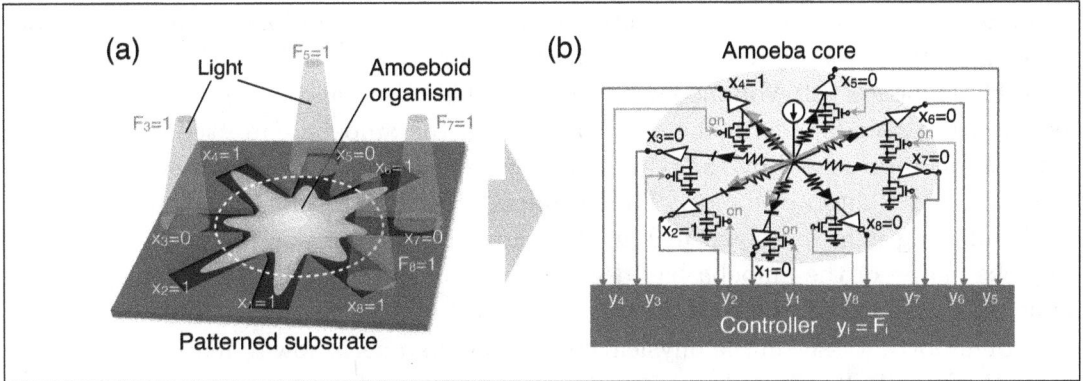

Figure 1: Basic concept of (a) amoeba-based computer and (b) electronic amoeba.

tiotemporal dynamics of the amoeba organism, called amoeba core. The amoeba core has many pseudopod units with a star topology and each pseudopod includes a capacitor, a diode, and a field-effect transistor (FET) together with an inverter for a threshold device. The dynamic behaviors of the amoeboid organism are represented by current in the pseudopod units and a state variable is represented with the current in a unit; when current flows in the pseudopod i, then, $x_i = 1$, else $x_i = 0$. The current flow is externally controlled by turning on/off the FET. The current in each unit is converted to a voltage signal through the capacitor and the inverter. The motive force of the amoeba organism for searching is to maximize food intake avoiding the risk, whereas that of the electronic amoeba is to reach steady state where the Kirchhoff laws are satisfied. The controller in the electronic amoeba maps a problem to be solved and defines the interaction between the state variables of the amoeba core. The interaction in the electronic amoeba is defined by bounceback rule [5]. The controller always watches the state variables in the amoeba core, checks them with the bounceback rule, and sends feedback signals to FETs for flipping the variables that violate the rule. Hence, the amoeba core searches the state that is not refused by the controller.

Aono *et al.* clarified that the amoeboid organism could search the solution of CSP (constraint satisfaction problem) [10] and TSP (traveling salesman problem) [11]. After these demonstrations, they extracted an abstract model in terms of solution search from the amoeba's foraging behavior and found its efficient solution search for SAT (satisfiability problem) [5]. The number of iterations of the amoeba-inspired SAT algorithm was smaller than that of WalkSAT algorithm [5]. Their amoeba model consists of three basic properties in terms of solution search as follows. (1) Stretching the pseudopods unidirectionally with fluctuation. The electronic amoeba represents this behavior by unidirectional current through a diode in the pseudopod unit. Fluctuation is represented by random error in the FET switching using random bit sequence. (2) Volume conservation. This arises spatiotemporal correlation between the pseudopods in the deformation. In the amoeba core, the volume conservation is represented by the Kirchhoff's current law in the hub of a star-topology network. (3) Photo avoidance. The pseudopod in the amoeba organism is shrunk by irradiating light as shown in Fig. 1(a) and this property is used to control the amoeba shape in the amoeba-based computer. The current in the amoeba core is control by turning off the FET in the pseudopod unit.

An important issue in the physical computing systems is how to map the problem. Here Boolean SAT is considered, which is the problem of determining whether there exist a variable vector $x = (x_1, x_2, ..., x_N)$ that makes a given Boolean formula $f(x)$ true, $f(x) = 1$. In the case of the electronic amoeba, the constraint function is converted to âĂIJbounceback ruleâĂİ [5]. The bounceback rule $\{F_{i\_j}\}$ ($i = 1, 2, ..., N$ and $j = 0$ or 1) consists of sub-rules that identify the variables which make the constraint formula false. First, we introduce a redundant variable expression; X_{i_0} and X_{i_1} for x_i. If $(X_{i\_0}, X_{i\_1}) = (1, 0)$ then $x_i = 0$, else if $(X_{i\_0}, X_{i\_1}) = (0, 1)$ then $x_i = 1$. Originally this expression was introduced because the pseudopod of the amoeba organism can be controlled only from 1 to 0, but not from 0 to 1 [5]. The redundant expression is still very useful even in the electronic amoeba, because this makes the numbers of variables 0 and 1 the same when the system finds a solution. This means that the Hamming distance of the solution vector is always N independent on the instance. Thereby the bounceback rule also includes the sub-rule that avoids X_{i_0} and X_{i_1} are flipped to 0 at the same time, called INTRA. The main part of the bounceback rule is to flip the variables that make each clause in the given formula false. We call this sub-rule as INTER. This sub-rule is derived from a conjunctive normal form (CNF) of $f(x)$. To satisfy $f(x) = 1$, each clause in the CNF of $f(x)$ should be 1. The other part of the bounceback rule is CONTRA which avoid inconsistency between INTER in terms of satisfying INTRA. An example of the bounceback rule is summarized in Table 1 for the instance $f(x) = (x_1 + x_2 + x_3)(\overline{x_1} + \overline{x_2} + \overline{x_3})(\overline{x_1} + \overline{x_2} + x_3)$. In this study we consider 3-SAT in which each clause contains three laterals. Finally, the entire form of the bounceback rule for variable X_{i_j} is given by $F_{i\_j} = \text{INTER}_{i\_j} \cup \text{INTRA}_{i\_j} \cup \text{CONTRA}_{i\_j}$.

The bounceback rule for SAT can be systematically and easily derived from the constraint formula. The electronic logic circuit can be used to map the given constrain, since the rule is described with Boolean functions and the feedback signal is binary. As a result, compared to the Ising model [12], the electronic amoeba can map SAT simply and easily. It is not necessary to estimate and assign multiple value to the interactions between the variables as in the Ising machine.

$$f(x) = (x_1 + x_2 + x_3)(\overline{x_1} + \overline{x_2} + \overline{x_3})(\overline{x_1} + \overline{x_2} + x_3) = 1$$

$$\rightarrow \begin{cases} X_{1\_1} + X_{2\_1} + X_{3\_1} = 1 \\ X_{1\_0} + X_{2\_0} + X_{3\_0} = 1 \\ X_{1\_0} + X_{2\_0} + X_{3\_1} = 1 \end{cases}$$

Variable	INTRA	INTER	CONTRA
X_{1_0}	X_{1_1}	$X_{2\_0} \cdot X_{3\_0}$	$X_{1\_0} \cdot X_{2\_0} \cdot X_{1\_1} \cdot X_{2\_1}$ $+ X_{1\_0} \cdot X_{3\_0} \cdot X_{1\_1} \cdot X_{3\_1}$ $+ X_{1\_0} \cdot X_{3\_0} \cdot X_{1\_1} \cdot X_{3\_0}$
X_{1_1}	X_{1_0}	$X_{2\_1} \cdot X_{3\_1} + X_{2\_1} \cdot X_{3\_0}$	$X_{1\_1} \cdot X_{2\_1} \cdot X_{1\_0} \cdot X_{2\_0}$ $+ X_{1\_1} \cdot X_{2\_1} \cdot X_{1\_1} \cdot X_{2\_1}$ $+ X_{1\_1} \cdot X_{3\_1} \cdot X_{1\_0} \cdot X_{3\_0}$ $+ X_{1\_1} \cdot X_{3\_0} \cdot X_{1\_0} \cdot X_{3\_0}$
X_{2_0}	X_{2_1}	$X_{1\_0} \cdot X_{3\_0}$	$X_{2\_0} \cdot X_{3\_0} \cdot X_{2\_1} \cdot X_{3\_1}$ $+ X_{2\_0} \cdot X_{3\_0} \cdot X_{2\_1} \cdot X_{3\_0}$ $+ X_{1\_0} \cdot X_{2\_0} \cdot X_{1\_1} \cdot X_{2\_1}$
X_{2_1}	X_{2_0}	$X_{1\_1} \cdot X_{3\_1} + X_{1\_1} \cdot X_{3\_0}$	$X_{2\_1} \cdot X_{3\_1} \cdot X_{2\_0} \cdot X_{3\_0}$ $+ X_{2\_1} \cdot X_{3\_0} \cdot X_{2\_0} \cdot X_{3\_0}$ $+ X_{1\_1} \cdot X_{2\_1} \cdot X_{1\_1} \cdot X_{2\_1}$
X_{3_0}	X_{3_1}	$X_{1\_0} \cdot X_{2\_0} + X_{1\_1} \cdot X_{2\_1}$	$X_{2\_0} \cdot X_{3\_0} \cdot X_{2\_1} \cdot X_{3\_1}$ $+ X_{2\_0} \cdot X_{3\_0} \cdot X_{2\_1} \cdot X_{3\_0}$ $+ X_{1\_0} \cdot X_{3\_0} \cdot X_{1\_1} \cdot X_{3\_1}$ $+ X_{1\_0} \cdot X_{3\_0} \cdot X_{1\_1} \cdot X_{3\_0}$
X_{3_1}	X_{3_0}	$X_{1\_1} \cdot X_{2\_1}$	$X_{2\_1} \cdot X_{3\_1} \cdot X_{2\_0} \cdot X_{3\_0}$ $+ X_{1\_1} \cdot X_{3\_1} \cdot X_{1\_0} \cdot X_{3\_0}$

Table 1: Example of SAT constraint formula and its bounceback rule. CONTRA intentionally includes several redundant terms to show that the rule is systematically derived.

In solution search process, the controller reads out state variables at time t, $\boldsymbol{X}(t)$, and checks the bounceback rule $F_{i\_j}(\boldsymbol{X}(t))$ for all i and j. Here, $\boldsymbol{X}(t) = (X_{1\_0}(t), X_{1\_1}(t), X_{2\_0}(t), X_{2\_1}(t), ..., X_{N\_0}(t), X_{N\_1}(t))$. Then the bounceback signal $Y_{i\_j} = \overline{F}_{i\_j}(\boldsymbol{X}(t))$ is feedbacked to the FET in the amoeba core so that the state variable at time $t + \Delta t$ is flipped to follow Y_{i_j}, $Y_{i\_j} \rightarrow X_{i\_j}(t + \Delta t)$. It should be noted that, however, $X_{i\_j}(t + \Delta t)$ does not always follow Y_{i_j}, because of the long capacitance charging time and externally introduced error. When the system finds a solution, the bounceback rule does not flip the variables further and the dynamics of the system is lost. It is also found that Hamming distance of the solution variable vector is always N: the half of X_{i_j} take 1 and the other half take

0. Additionally, when the system finds a solution, the half of F_{i_j} hold 0 and the others hold 1, which stabilize X_{i_j} at 1 and 0, respectively.

3 Experiment

The electronic amoeba solving SAT was implemented by integrating the commercial discrete electronic devices together with a microcomputer for the controller as shown in Fig. 2(a). The capacitance C and resistance R in each pseudopod were 10 nF and 100 Ω, respectively. The source current in the hub was 10 μA. The number of variables N was 12, then the number of the redundant variables $2N$ was 24. The

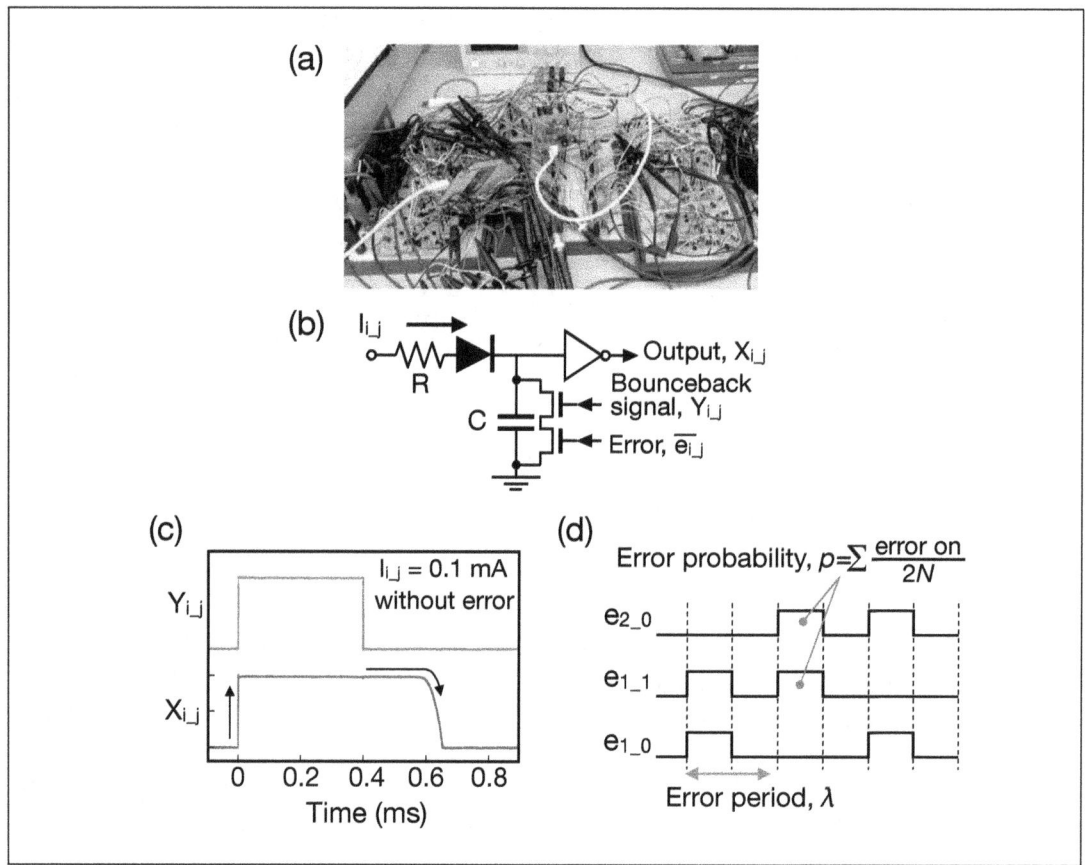

Figure 2: (a) Photograph of the fabricated electronic amoeba, (b) circuit diagram of a pseudopod unit, (c) example of the response of the pseudopod to bounceback signal, and (d) typical error signal waveforms.

bounceback rule was programmed in the microcomputer. The clock frequency of the controller was 84 MHz. For fluctuation, we used random bit sequences generated by Xorshift method. The error signal was randomly given to each pseudopod via a FET in series with the FET for the bounceback control as shown in Fig. 2(b). Figure 2(c) shows a measured response of the pseudopod unit X_{i_j} to the bounceback signal Y_{i_j}. The output X_{i_j} immediately rose when Y_{i_j} became high, since the capacitance discharging rapidly occurred when the FET turned on. On the other hand, when Y_{i_j} became low, X_{i_j} showed a long delay. This was because the capacitance charging took long time due to small injected current from the hub. The long charging time biased the state variable to 1. When the current of the current source in the hub was increased, this delay was linearly shortened. The error signal was given to make the variable 0 regardless the bounceback signal. Here we also investigated the effect of the two properties of the error on the solution search as shown in Fig. 2(d). Error probability p means how many pseudopod units receive error signal at an error period. Error period λ is defined by the sum of an error bit width and average time interval between the neighboring error bits. Increasing λ, both the error bit width and bit interval are increased. The state variables of the electronic amoeba were read out from the output voltages of the inverters in the end of the pseudopod units. The outputs were recorded using a 24-channel oscilloscope.

4 Results and discussion

Figure 3(a) shows the time evolutions of the variables in searching process of a 12-variable SAT. The random error of $p = 0.65$ and $\lambda = 0.6$ ms was used. In this plot selected four variables are shown. Starting from the initial state $X_{i\_j} = 1$ for all i and j, each variable irregularly flipped between 0 and 1, and sometimes showed oscillatory behaviors. When the error signals were imposed to the pseudopod units, the state of the amoeba core was forcibly changed. After imposing the errors several times, the system reached a steady state and became stable. We confirmed that the values of the variables in the stable state always satisfied the constraint, that is, the amoeba core found a global solution. This was simply because the bounceback rule continued to flip one of the state variables at least until the all variables satisfied the constraint. The oscillatory state in the searching process suggested that the amoeba core could not solve the inconsistency between the variables and the system fell in a localized state. In the worst case, the electronic amoeba cannot escape from the localized state and the oscillatory behavior maintains. However, this suggests that we can identify when the state of the amoeba core should be forcibly changed. This property is obviously different from the Ising machine where it is not easy to

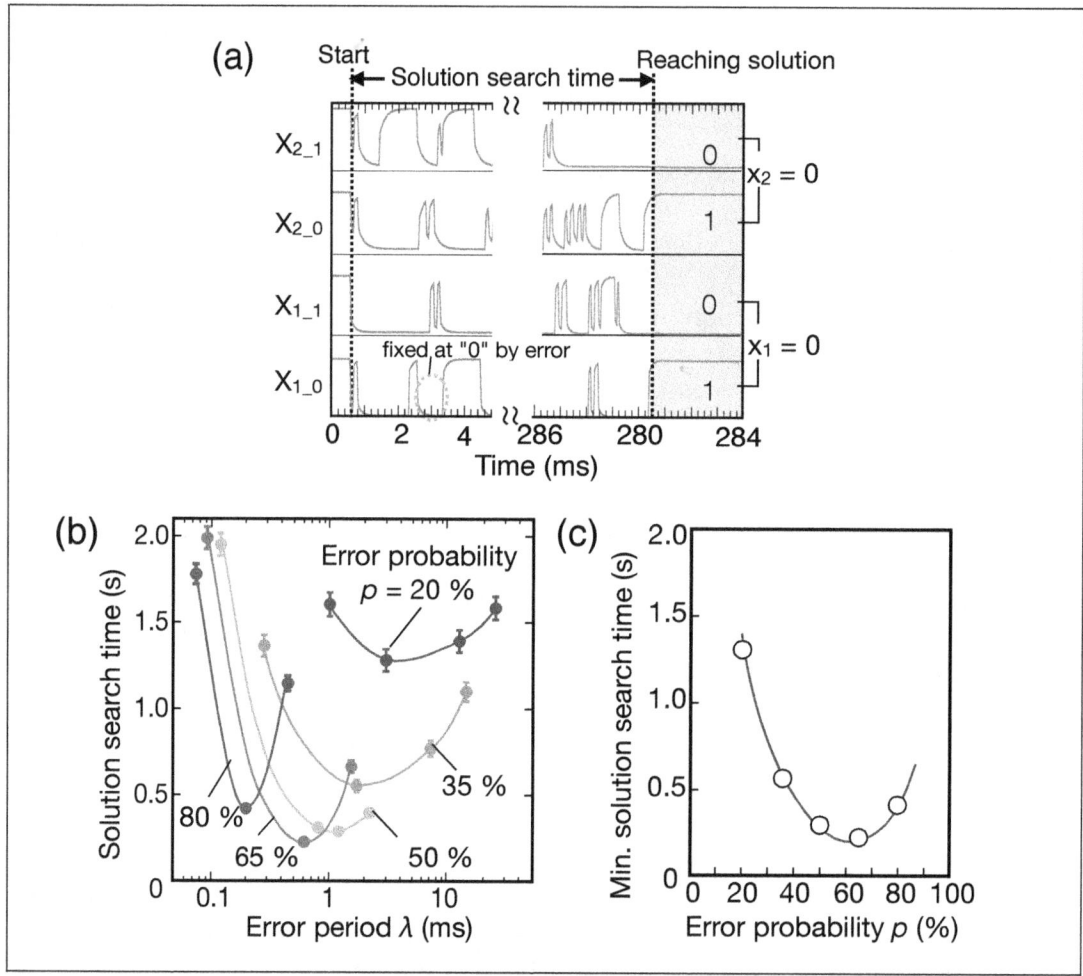

Figure 3: (a) Time evolution of state variables in the amoeba core solving 12-variable SAT. $p = 65\%$, $\lambda = 0.6$ ms, only selected four variables are shown. (b) Evaluated solution search time as a function of error period for various error probabilities, and (c) minimum solution searching time as a function of error probability.

distinguish the global minimum from the local minimum. The random error in the electronic amoeba was found to help the system escaping from the localized state, similar to the annealing in the Ising machine.

The solution search of the electronic amoeba was expected to depend on the error property. Figure 3(b) shows the evaluated SAT solution search time as a function of the error period for various error probability p. Each data point was obtained by solving 25 instances 30 times. It was clearly found that the error property affected the solution search efficiency. This behavior was not seen in the solution search by the amoeba algorithm executed by a conventional computer. Each curve in the plot had a minimum value, showing the existence of the optimal error period λ that gave the minimum searching time. The minimum solution searching time also depended on the error probability p as shown in Fig. 3(c) and this indicated that the optimal error probability also existed. In addition, the optimal error period changed depending on the error probability. The smallest solution search time of 0.28 s was obtained with $p = 65\%$ and $\lambda = 0.6$ ms in the present system.

To understand the obtained behaviors, we considered the time scale of the amoeba core dynamics and the distance in the variable space that the amoeba core was moved by the errors in each error period. The circuit configuration of the pseudopod unit in Fig. 3(b) indicates that an error signal fixes a variable at 0. Thus the error signals effectively decreased the variable space where the amoeba could move around. When the error probability is p, the number of the fixed variables is approximately given by $p \cdot 2N$. Then, Hamming distance from the origin of the variable space, H, follows the condition $H \leq (1-p) \cdot 2N$. Considering the variables are biased to 1 owing to the larger capacitance charging time than the discharging time, H tends to be $(1-p) \cdot 2N$. Because the Hamming distance of the solution vector is always N when the redundant variable expression is used, the state of the amoeba core is found to be close to the solution vector when $p = 0.5$. This results in the small solution search time around $p = 50\%$, which reasonably explain the obtained result.

The observed error-period dependence was attributed to the balance between the capacitance charging time and the error period. The time scale of the amoeba motion can be evaluated by the capacitor charging time in each pseudopod unit, which is approximately given by $\tau_c \sim C \cdot V_{th}/(2N \cdot (1-p) \cdot I)$, where C is capacitance, V_{th} is the threshold voltage of the inverter in the pseudopod unit, and I is the current from the current source. When p is assumed to be 0.5, τ_c becomes 0.2 ms. The discharging time of the capacitance τ_d is enough small compared to the charging time, $\tau_c \gg \tau_d$, since the discharging immediately occurs when the FET turns on. When the error period is much shorter than the charging time, $\tau_c \gg \lambda$, the amoeba core cannot search around whole the variable space, because it does not have enough time to change the state in each error period. This makes the amoeba core frequently hop from point to point in the variable space and then the amoeba core likely misses the solution even when the solution exists near the hopping point. On the other hand, when the error period is much longer than the charging time, $\tau_c \ll \lambda$, the amoeba core spends extra time when it falls in a localized state. This delays the amoeba core to reach the solution. Therefore, it is found that $\tau_c \approx \lambda$ is necessary and sufficient for searching around the variable space. The error period achieving the minimum search time in Fig. 3(b) was around 1 ms, which was reasonably matched with evaluated τ_c of 0.2 ms.

The discussion above suggests that the decreasing the charging time by reducing the capacitance or increasing the current supplied to the hub decreases the solution search time. From viewpoints of the low power consumption, decreasing the capacitance is appropriate. However, in this study, we used rather large capacitance to make the dynamics of the amoeba core slow so that the controller with 84 MHz clock could follow in detail the amoeba core behavior operating asynchronously. To compare the speed and power consumption with the Ising machines, it is also necessary to increase the number of the variables up to thousands [1, 2, 13] and this is now under investigation. The obvious advantage is that the number of variables for mapping SAT is $2N$ in the electronic amoeba, whereas N^2 order in the Ising machine owing to the lattice configuration of the spatial arrangement of the variables. Furthermore, the SAT instance can be easily mapped on the electronic amoeba by Boolean logic gates, whereas Ising model needs to translate the instance to a set of real numbers and assign it to each interaction of the variables. The advantages above indicate the simple and small implementation of the electronic amoeba with low power consumption compared to the Ising machines.

A separate study indicates that the number of iterations in the amoeba-inspired SAT algorithm increases less than exponentially when N is increased, because the several variables are updated simultaneously in each iteration step. In the case of the electronic amoeba, a short period of the oscillations in the time evolution of the variables effectively corresponds to an iteration. A period of the variable oscillation is determined by capacitance charging time in the pseudopod unit and the clock frequency of the controller is high enough compared to the oscillation period. Then the controller can flip several variables in each period of the oscillation even though the bounceback rule is evaluated in a serial manner. Thus the solution search time is expected to increase less than exponentially as a function of N in the electronic amoeba. Further study is necessary to confirm this point. It is noted that, in our system, the controller requires computation power to evaluate the bounceback rule, however the solution cannot be found without the amoeba core; if the amoeba core completely follows the bounceback rule, the system cannot find a solution.

5 Application to autonomous robot walking

The feasibility of the electronic amoeba was investigated by the demonstration of the autonomous control of the robot . We implemented the electronic amoeba to a commercially available four-legged robot as shown in Fig. 4(a) and examined its autonomous walking. The robot was designed to walk straight without programing the leg maneuver. Then the robot was only provided with the constraints to avoid falling and straying. This robot is referred as amoeba robot hereafter. The electronic amoeba was expected to find an appropriate leg maneuver according to its posture in each step. Each leg had 3 × 2 motion states as shown in Fig. 4(b). The posture information was obtained from the touch sensor in each foot together with the previous state variables. Then the electronic amoeba searched 24 state variables (4 legs × 6 states) under the constraints described with the bounceback rule. In this experiment, both the amoeba core and the controller were implemented in the same microcomputer for simple and compact implementation at this time. The clock frequency of the microcomputer was 16 MHz.

First we obtained information of the unfavorable actions for the robot walking. If these actions were described as bounceback rule, the amoeba robot was expected to search favorable actions by avoiding unfavorable actions depending on the environment and its posture at each moment. We found 13 rules for the autonomous walking by considering the body balance and careful observation of the robot behaviors under various test commands. Figure 5 shows the three major motions resulting in failure in walking straight. The obtained bounceback rule is summarized in Table 2. The rule also includes a few codes suggesting the objective of the robot. A global solution in this case corresponds to the balanced posture. Information of the environment and the posture was obtained from the leg position and the touch sensor on each foot. The error signal was also randomly imposed to the amoeba robot in the solution search process, which was necessary to break the steady state and to take

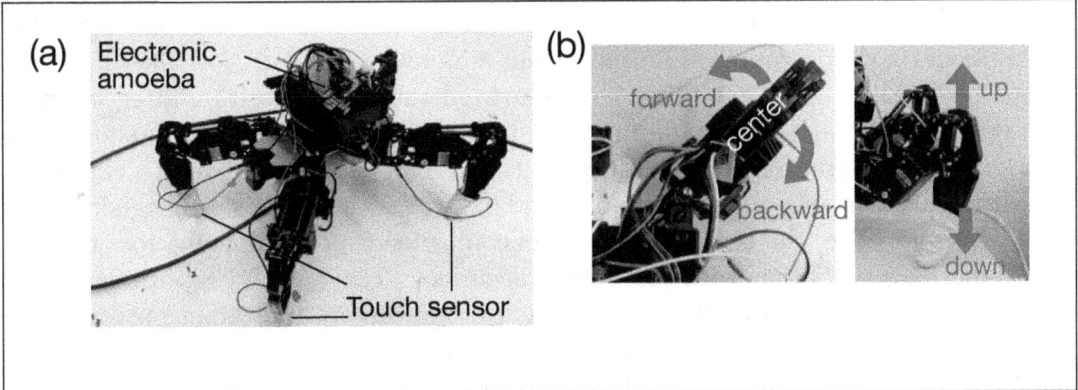

Figure 4: (a) Photograph of a four-legged amoeba robot and (b) possible leg state for walking control.

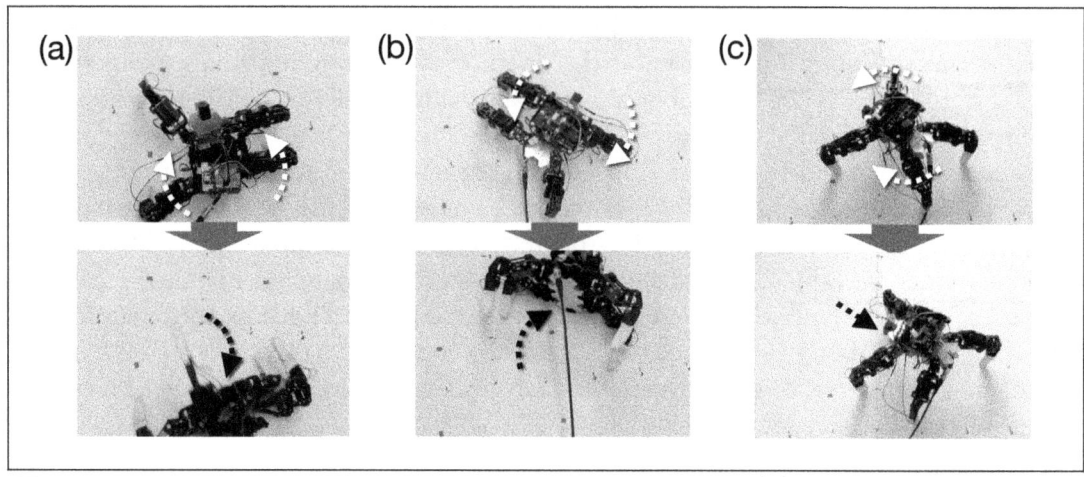

Figure 5: Typical motion patterns avoided for walking: (a) falling backward, (b) falling forward, and (c) moving backward.

the next walking step. In general case, the bounceback rule for autonomous walking might be deduced from the body balance without empirical testing. A potential way to improve the rule without observation is reinforcement learning through experience and this approach is now under consideration.

Figures 6(a-e) and 6(f-j) show the movements of the robot with a controller programmed with conventional walking patterns and the amoeba robot finding a leg maneuver in realtime, respectively. The walking target was set at 80 cm from the start position as indicated with arrows in Figs. 6(a) and 6(f). The amoeba robot could reach the target, although its track was fluctuated compared to the conventional walking robot. The movement was awkward and slow like a baby's crawl, however, it never fell. The deviation from the target position after walking was evaluated by the error angle from the target direction. Figure 6(k) shows the histogram of the error angle taken after 100 trials. The histogram showed a Gaussian like distribution having a peak around 0° and 68 % of the trials was converged within ±30° deviation from the center. The results confirmed that the amoeba robot could reach the target position with reasonable accuracy. The amoeba robot consumed extra time to reach the target position: the average time to the target position was 83 s, whereas that of the robot with the conventional walking program was 25 s. The average thinking time in one step, corresponding to the time to find a solution, was estimated to be 1.5 s.

The awkward walking of the amoeba robot was attributed to not only the fluc-

1	If number of legs with sensor on < 2, then prohibit legs up.	
2	If one of forelegs is twisted back, then prohibit another from moving back.	
3	If one of hind legs is twisted forward, then prohibit another from moving forward.	
4	If $Leg(i)$ is twisted forward, prohibit leg in diagonal position from moving excepting forward.	
5	If $Leg(i)$ is twisted back, prohibit leg in diagonal position from moving excepting back.	
6	If $Sensor(i)$ is on, then prohibit $Leg(i)$ from moving forward + down, and permit $Leg(i)$ moving back + down.	
7	If $Sensor(i)$ is off, then prohibit $Leg(i)$ from moving back + up and permit $Leg(i)$ moving forward + up.	
8	If $Leg(i)$ is down and $Sensor(i)$ is off, then permit $Leg(i)$ moving laterally.	
9	If $Leg(i)$ is up and $Sensor(i)$ is off, then permit $Leg(i)$ down and keeping lateral position.	
10	If $Leg(i)$ is up and $Sensor(i)$ is off, then permit $Leg(i)$ down and laterally moving to another side.	
11	If $Leg(i)$ dis own and $Sensor(i)$ is on, then permit $Leg(i)$ up keeping lateral position.	
12	If $Leg(i)$ is down and $Sensor(i)$ is on, then permit $Leg(i)$ up and laterally moving to another side.	
13	If $Leg(i)$ is up and $Sensor(i)$ is on, then permit $Leg(i)$ down keeping lateral position.	

Table 2: Bounceback rule for autonomous waking of a four-legged amoeba robot.

tuation of the walking direction but also the variation of the thinking times in step by step. The thinking time mainly arose from the solution search. Increase of the searching speed of the electronic amoeba would contribute to the fast and smooth walk. High speed solution search is also indispensable for the safe recovery when the unexpected accident occurs in walking. The delay in each action makes it difficult to recover from the sudden change of the body balance, resulting in falling down.

To clarify the feasibility of optimization-based autonomous walking, the amoeba robot was examined walking when there were obstacles preventing its walking. An expectation of the autonomous robot walking based on the optimization could keep walking by finding appropriate leg maneuvers even when unexpected things hap-

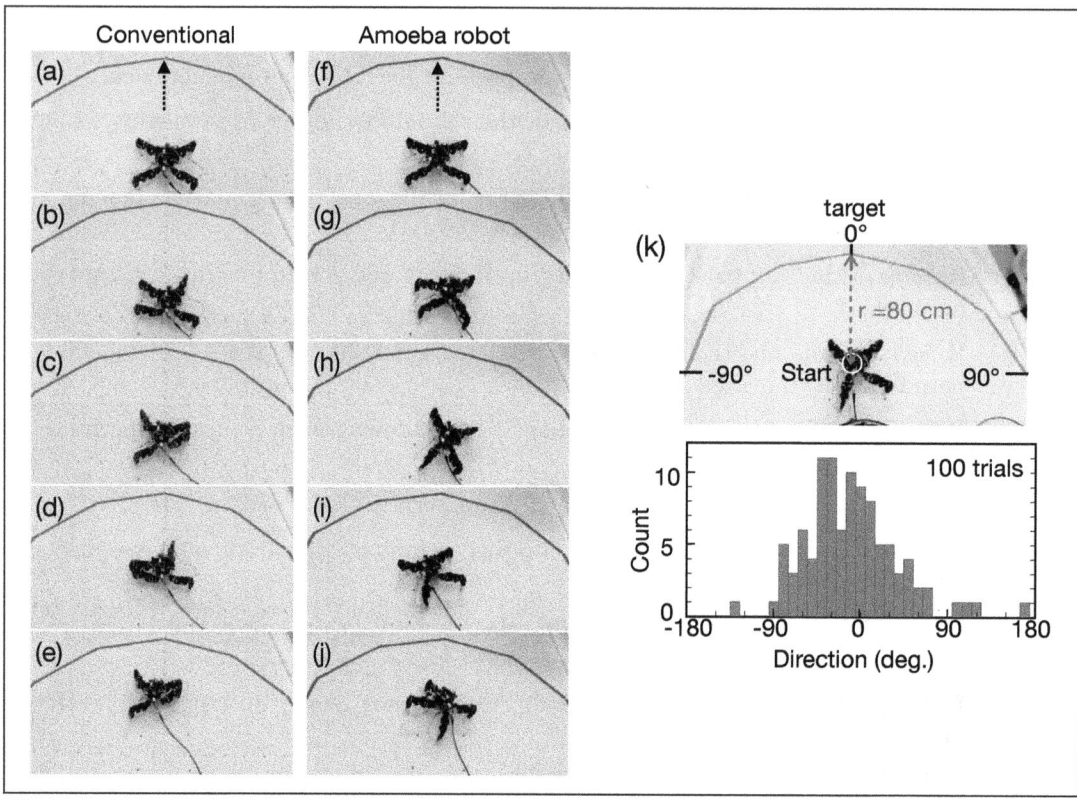

Figure 6: Snapshots of movements of the four-legged robots: (a)-(e) conventional walk programmed with standard leg maneuver and (f)-(j) amoeba robot with autonomous walking control. Red line is equidistance of 80 cm from the start position, and (k) histogram of walking direction of amoeba robot at 80 cm from start point with a map indicating direction angles. 0° corresponds to arrival at target point.

pened. As the obstacle, the two poles with pedestals were set in front of the robot as shown in Fig. 7. No information of the obstacles was given to the robot beforehand. In the case of the conventional walking program, the robot repeated the same motions and never escaped from the obstacles (Figs. 7(a-c)). On the other hand, the amoeba robot could go over the obstacles without falling after several trials and errors as shown in Figs. 7(d-f). The result indicated that the optimization-based autonomous control arose the ability to find a way to deal with unknown events and unexpected situations.

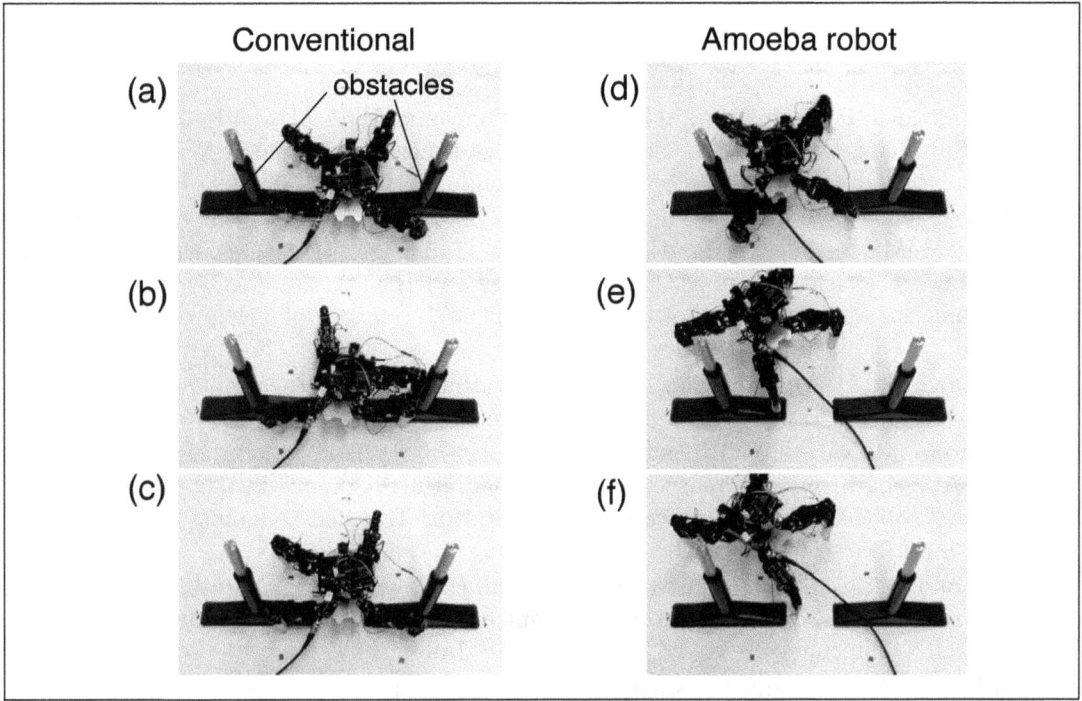

Figure 7: Snapshots of movement of four-legged robot with obstacles on its lane: (a)-(c) conventional walking program and (d)-(f) autonomous walking control by an electronic amoeba.

6 Conclusion

We have developed an amoeba-inspired electronic computing system, electronic amoeba. This system searched for a solution to a combinational optimization problem, as inspired by foraging behavior of a single-celled amoeboid organism that maximizes its food intake under given constraints. We electronically implemented the system and demonstrated its solution search capability for solving Boolean satisfiability problem (SAT), where the constraint function was mapped on the system using bounceback rule. We applied the electronic amoeba to autonomous walking control of a four-legged robot. It was demonstrated that the electronic amoeba successively searched a combination of the leg joint states to satisfy the objective of moving straight depending on the state of the robot and it also had a ability to deal with unknown situations. Combination of the amoeba-inspired solution search system with the machine learning is expected to achieve smart autonomous control of the robot and other machines.

References

[1] https://www.dwavesys.com/home

[2] M. Yamaoka, C. Yoshimura, M Hayashi, T. Okuyama, H. Aoki, and H. Mizuno, "A 20k-Spin Ising Chip to Solve Combinatorial Optimization Problems With CMOS Annealing," IEEE Journal of Solid-State Circuits, vol. 51, pp.303-309, Jan. 2016.

[3] T. Inagaki, Y. Haribara, K. Igarashi, T. Sonobe, S. Tamate, T. Honjo, A. Marandi, P. L. McMahon, T. Umeki, K. Enbutsu, O. Tadanaga, H. Takenouchi, K. Aihara, K. Kawarabayashi, K. Inoue, S. Utsunomiya, and H. Takesue, "A coherent Ising machine for 2000-node optimization problems," Science vol.354, pp. 603-606, Oct. 2016.

[4] T. Nakagaki, H. Yamada, and A. Tóth, "Intelligence: Maze-solving by an amoeboid organism," Nature vol.407, p.470, Sep. 2000.

[5] M. Aono, M. Naruse, S.-J. Kim, M. Wakayabayashi, H. Hori, M. Ohtsu, and M. Hara, "Amoeba-Inspired Nanoarchitectonic Computing: Solving Intractable Computational Problems Using Nanoscale Photoexcitation Transfer Dynamics," Langmuir vol.29, pp.7557-7564, Apr. 2013.

[6] M. Aono, S. Kasai, S.-J. Kim, M. Wakabayashi, H. Miwa, and M. Naruse, "Amoeba-inspired nanoarchitectonic computing implemented using electrical Brownian ratchets," Nanotechnology, vol.26, art.no.234001, Jun. 2015.

[7] S. Kasai, M. Aono, and M. Naruse, "Amoeba-inspired computing architecture implemented using charge dynamics in parallel capacitance network," Appl. Phys. Lett. vol.103, art.no.163703, Oct. 2013.

[8] K. Saito, N. Suefuji, S. Kasai, and M. Aono, "Amoeba-inspired electronic solution-searching system and its application to finding walking maneuver of a multi-legged robot," presented at 48th IEEE International Symposium on Multiple-Valued Logic (ISMVL2018), Linz, Austria, May 16-18, 2018.

[9] G. A. Bakey, Autonomous Robots: From Biological Inspiration to Implementation and Control, ed. C. Arkin, MIT press, 2005.

[10] M. Aono, M. Hara, and K. Aihara, "Amoeba-based Neurocomputing with Chaotic Dynamics," Commun. ACM vol.50, pp.69-72, Sep. 2007.

[11] L. Zhu, M. Aono, S.-J. Kim, and M. Hara, "Amoeba-based computing for Traveling salesman problem: Long-term correlations between spatially separated individual cells of Physarum polycephalum", BioSystems vol.112, pp.1-10, Apr. 2013.

[12] M. Ercsey-Ravasz and Z. Toroczkai, "Optimization hardness as transient chaos in an analog approach to constraint satisfaction," Nature Phys. vol.7, pp.966–970, Dec. 2011.

[13] H, Takesue, T. Inagaki, and I Inaba, "Solving large-scale optimization problems with coherent Ising machine," Proceedings of 2017 Conference on Lasers and Electro-Optics Pacific Rim (CLEO-PR), S1561, Jul. 31 - Aug. 4, 2017, Singapore. DOI: 10.1109/CLEOPR.2017.8118781.

Characterizing Parallel Multipliers for Detecting Hardware Trojans

Akira Ito, Rei Ueno, Naofumi Homma, and Takafumi Aoki
Tohoku University, Japan
{ito, ueno, homma}@riec.tohoku.ac.jp

Abstract

This paper presents a new analysis method for estimating the detectability of a hardware trojan (HT) that causes a path delay fault (PDF) to parallel multipliers. The proposed method characterizes a parallel multiplier with the average delay of all paths in a multiplier. We show that the average delay, which is determined by its multiplier structure, has a relation to the HT detectability. The validity of our method is evaluated by an experiment using Monte Carlo tests that measure the detection probabilities of HTs inserted into typical multipliers, and multiple regression analysis. In addition, we demonstrate how the amounts of inserted delay have impacts on the HT detectability. The result shows that, given an inserted delay amount and a multiplier structure, our analysis is useful for estimating the detectability.

Keywords: Hardware trojans, Arithmetic algorithms, Multipliers, Path delay faults

1 Introduction

Hardware Trojan (HT) threats have been a topic of significant interest in hardware security research. An HT is a hardware-oriented backdoor that can be inserted into cryptographic hardware to retrieve secret information. Modern IC chips including cryptographic hardware are manufactured by work division among many parties (e.g., fabless companies, design houses, IP venders, and semiconductor foundries), which might not be always trustworthy. In other words, a malicious party may insert an HT into cryptographic hardware to retrieve secret information from the chip users and/or their clients.

There are many previous works on HT insertion and detection (i.e., countermeasures). In earlier related works, many HTs that employ circuit function modification have been investigated. For example, one HT modifies the cryptographic datapath to output the secret key when the HT is triggered. Another HT described in [1] modifies the datapath to cause one or more faults in the cryptographic operations to extract the secret key by using a type of fault-based cryptanalysis called differential fault analysis [2]. The abovementioned HTs consist of trigger and payload units. The trigger unit activates the payload unit only when the cryptographic hardware has specific input values. Since the trigger values are very limited and unknown to chip testers, it is difficult to detect them during Monte-Carlo-based chip tests. Furthermore, it is impractical to perform an exhaustive test for cryptographic hardware because the primary inputs are usually longer than 128 bits. However, as these HTs modify the chip geometry and/or explicitly add extra specific blocks/paths, many countermeasures against them exploits the differences between the malicious chips and the Trojan-free golden models. For example, it is known to check products with scanning electron microscope (SEM) images of manufactured chips, or compare the side channel information of chips with the golden models. Similarly, an HT that modifies the dopant polarities of cells to cause stuck-at faults intentionally [3] can even be detected by SEM imaging with feasible additional procedures.

In contrast, a new type of HT called a Path Delay HT (PDHT), presented in [4], causes faults (i.e., bugs) in multipliers seemingly without modifying the circuit functions at the logic or cell level (i.e., without additional trigger/payload units or stuck-at faults). While integer multiplication is one of the major operations in public key cryptography (e.g., RSA [5] and elliptic curve cryptography [6]), bugs in multipliers in public key cryptographic hardware can be exploited by attackers to retrieve secret keys [7]. The approach used by a PDHT involves finding a rarely sensitized path called a rare path (RP) in a multiplier[1] and replacing gates along the RP with the same functional gates with larger delays such that the RP delay becomes larger than the critical delay. The output of a PDHT-inserted multiplier is buggy due to the setup time violation only when the inputs are the specific values required to sensitize the RP. Note that it is quite difficult to detect PDHTs by performing Monte Carlo tests because RPs are sensitized by few inputs. It is also difficult to detect PDHTs even with SEM images because PDHTs are inserted without additional units or faulty cells. Thus, PDHTs are considered to pose a serious threat to information system security.

On the other hand, while there are many hardware algorithms for parallel multiplication [8], the generality and applicability of PDHTs to such various multipliers are unclear. Although RPs are fully sensitized only by specific inputs, the extra delay added to an RP also influences other paths. If this influence is non-negligible,

---

[1] A path is sensitized if all of the gates on the path switch in a clock cycle. Note that gate switching due to glitch effects (or dynamic hazard) is not discussed in this paper, as in [4].

a PDHT-inserted multiplier can generate faulty output values even when the RP is not fully sensitized. In other words, PDHTs can be detected during Monte Carlo tests in such cases. The influence depends on the characteristics of the RP in the multiplier related to its hardware algorithm. While a method of suppressing the detectability (i.e., influence) of PDHT insertion was proposed in [4], the detectability of the inserted PDHT was evaluated through only one multiplication algorithm. Accordingly, the characteristics of RPs in multipliers, namely, the extent to which hardware algorithms for multiplication impact the insertion of PDHTs with low detection probabilities in Monte Carlo tests, should be studied to develop PDHT countermeasures.

In this paper, we present analyses of some typical multipliers from the viewpoints of RP characteristics and PDHT insertion/detection probability. We discuss how the delay added to an RP affects other paths. In particular, we present analyses of the statistical properties of the switching probability and number of gates along an RP. As a result, we demonstrate that the detectability is closely related to the difference between the critical delay and average delay of all of the paths in the multiplier in addition to the first-order statistical moment of the switching probability and the number of gates along the RP. We validate our argument by presenting the results of experimental PDHT insertion into some typical multipliers of different bit lengths. Here, we attempt various amounts of inserted delays in order to clarify the effect of delay insertion. Consequently, we demonstrate that multipliers based on redundant binary trees provide greater detectability than other multipliers.

2 Path Delay HT

The basic approach used by a PDHT involves modifying a path in a multiplier and letting the multiplier output faulty values when specific values that sensitize the path are input. The faulty outputs of such PDHT-inserted multipliers enable the attackers who inserted the PDHT to retrieve the secret key based on the bug attack [7]. While the conventional HTs are based on circuit function modification or stuck-at faults caused by added/modified blocks, paths, and gates, PDHTs are based on path delay faults [9]. In the path delay fault model, the path delay becomes longer than the clock period due to the long delays of gates along the path. The PDHT intentionally causes such path delay faults by replacing the gates along the RP with gates with the same function but longer delays.

An RP is a path sensitized with an extremely low probability. The RP can be sensitized only by attackers who know specific values, while it cannot be sensitized and detected by Monte Carlo tests. In addition, as PDHTs employ only valid and

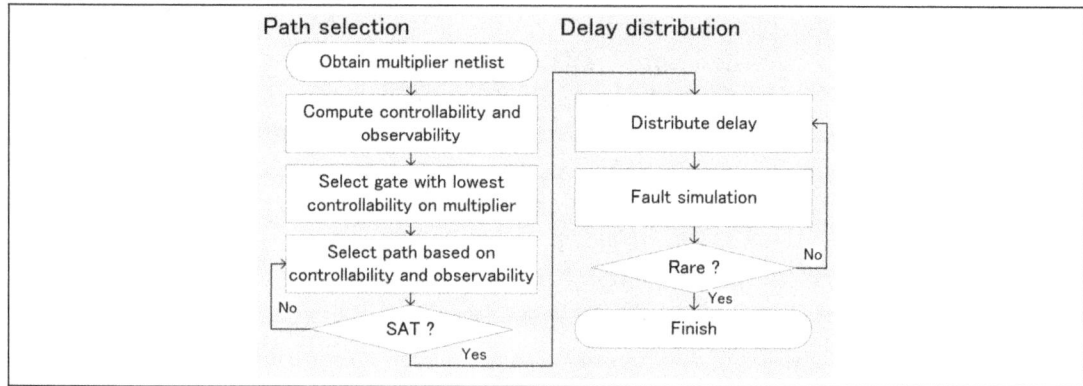

Figure 1: Flowchart of PDHT insertion.

correct cells (i.e., gates), it is difficult to detect PDHTs via conventional HT detection methods, which employ reverse-engineering techniques to find added/modified suspicious blocks, paths, and cells (even at the dopant level).

Figure 1 shows the flowchart of PDHT insertion presented in [4], which consists of two phases: RP selection and delay distribution. The RP selection phase finds an RP. Since the detectability of a PDHT depends on the probability of activating the RP (i.e., the total switching probability of the gates throughout the RP), the RP selection exploits two values related to switching probability: controllability and observability [10]. Hence, in the RP selection phase, the controllabilities and observabilities are first calculated.

Controllability is the probability of 0 or 1 on a wire, which connects two gates, if the primary inputs of the circuit are uniformly distributed. Observability is the probability that a value on a wire affects the primary output of the circuit.

Let $C_0(m)$ and $C_1(m)$ be the controllabilities of 0 and 1 on wire m, respectively, and let $B_0(m)$ and $B_1(m)$ be the observabilities of 0 and 1 on wire m, respectively. For example, let us consider a two-input AND gate. Let i and j be the input wires to the AND gate and the value of each wire be independent. Let k be the output signal. In this case, some controllabilities and observabilities are as follows:

$$C_0(k) = 1 - C_1(i)C_1(j), \quad (1)$$
$$C_1(k) = C_1(i)C_1(j), \quad (2)$$
$$B_0(i) = B_0(k)\left[\frac{C_1(j) - C_1(k)}{C_0(i)}\right], \quad (3)$$
$$B_1(i) = B_1(k)\frac{C_1(k)}{C_1(i)}, \quad (4)$$

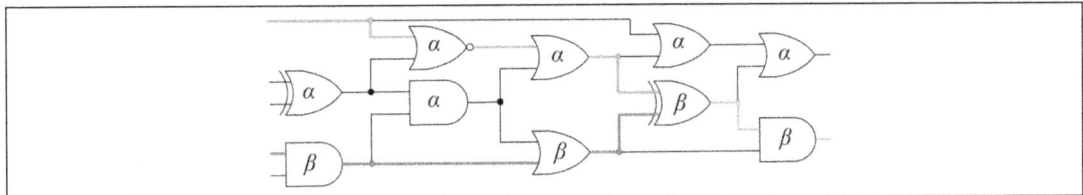

Figure 2: Example circuit.

where $C_0(k)$ and $C_1(k)$ are represented by the controllabilities of inputs i and j. The above equations indicate that the path activation probabilities (i.e., PDHT detectability) can be roughly calculated based on the controllabilities of the primary inputs. For RP selection, the controllabilities and observabilities are asymptotically calculated using a Monte Carlo method with logic simulation due to the difficulty of calculating their exact values for large multipliers deductively.

An RP is then selected according to the calculated controllabilities and observabilities. When an output wire of a gate has low controllability, it can be said to have low switching probability. A wire with low observability has a small influence on the primary output. Therefore, the RP is selected by identifying a series of wires (i.e., path) with lowest possible controllability and observability. More precisely, the wire with the lowest controllability is selected and is then extended to a primary input and an output with lowest possible controllabilities and observabilities. On the other hand, since the selected path is not always sensitizable, a SAT solver is used to check whether the path can be sensitized. Thus, an RP with a lowest possible probability of activation can be selected. In the delay distribution phase, a delay is added to each gate along the selected RP to minimize the probability of setup time violations (i.e., PDHT detection) during Monte Carlo testing. Since the number of paths in the multiplier increases exponentially as the gate depth increases, it is difficult to determine how an added delay affects other paths. Therefore, in [4], a genetic algorithm was used to determine how to add a delay to each gate along an RP.

3 Analysis of RP characteristics

In this section, we present analyses of RP characteristics and discuss their relations to PDHT detectability. We assume here that a PDHT can be detected if a setup time violation (i.e., an erroneous output) occurs during a Monte Carlo test with a clock period equal to the critical delay. Our analysis method focuses on the controllability and number of gates along the RP. When a setup time violation occurs, a primary input vector should partially sensitize the RP. In other words, it is important to analyze the number of switched gates along the RP when the delay is longer than the critical delay. To clarify the importance of the signal controllability and number of gates along the RP, we first consider an example circuit into which a PDHT has been inserted. This circuit is depicted in Fig. 2, where α is the original gate delay and $\beta(=\alpha+\gamma)$ is the modified gate delay obtained by adding an extra delay γ. For simplicity we assume that α is roughly the same regardless of the gate function. In Fig. 2, the RP is denoted by the red and green lines, and each gate along the RP has a delay of β. A path including some of the gates along the RP is denoted by the blue and green lines, and this path can be sensitized by Monte Carlo testing. The two paths share the partial path denoted by the green line.

In this case, when the path is sensitized, a setup time violation occurs if $2\alpha + 2\beta > d_{CP}$, where $d_{CP}(\approx 5\alpha)$ is the critical delay. More generally, let x and y be the numbers of switched gates with delays of α and β, respectively, along an sensitized path. Note that $x + y$ should be less than the number of gates along the critical path (and $x + y \leq 5$ in this case). Since the delay of the path is given by $x\alpha + y\beta = (x+y)\alpha + y\gamma$, the added delay $y\gamma$ mainly determines whether or not a setup time violation occurs. The added unit delay γ is smaller if the number of gates along the RP is larger. In addition, y is smaller if the controllability of signals along the RP is smaller. Thus, the controllability and number of gates along the RP have essential roles in determining the possibility of PDHT insertion/detection.

Let us then consider the general case. Let P_{RP} and P be an RP and a path that can be sensitized by Monte Carlo testing, respectively. In addition, let d_g and d'_g denote the delays of gate g before and after delay insertion, respectively[2]. A path is defined as a set of gates. For example, gate g in path P is denoted as $g \in P$ and the delay of P is denoted as d_P. The delay of P after delay insertion is

$$d'_P = d_P + \sum_{g \in P \cap P_{RP}} (d'_g - d_g), \qquad (5)$$

where $P \cap P_{RP}$ denotes a set of switched gates on both P and P_{RP} (i.e., sensitized gates on RP). The condition that causes a setup time violation by activating P is $d'_P > d_{CP}$, where d_{CP} denotes the critical path delay. For simplicity, the added delay is assumed to be uniformly distributed over the gates[3]. Then, Eq. (5) can be rewritten as

---

[2]In general, each gate delay differs owing to whether a rising or falling transition happends in the gate and which input port is active. The following discussion can also applied to such a more precise model.

[3]This assumption was made because it enabled the effects of switched gates along the RP to be estimated. In [4], it was demonstrated that an optimally distributed delay based on the genetic algorithm can reduce the detectability to at most one-fourth of its original value in comparison with

$$d'_P = d_P + |P \cap P_{RP}| \frac{d'_{RP} - d_{RP}}{|P_{RP}|}, \qquad (6)$$

where $|P_{RP}|$ and $|P \cap P_{RP}|$ denote the number of gates along P_{RP} and switched gates, respectively. In addtion, d_{RP} and d'_{RP} are the RP delays before and after delay insertion, respectively. Thus, the $|P \cap P_{RP}|$ condition that represents PDHT detection by activating P is

$$|P \cap P_{RP}| > |P_{RP}| \frac{d_{CP} - d_P}{d'_{RP} - d_{RP}}. \qquad (7)$$

Since P of Eq. (7) represents an arbitrary path in a multiplier, we analyze the statistical properties of the right hand side (RHS) of Eq. (7). Let $E[f(P)]$ be the first-order statistical moment of a function $f(P)$ (i.e., the average value of $f(P)$). The average value of the RHS of Eq. (7) can be written as

$$E\left[|P_{RP}| \frac{d_{CP} - d_P}{d'_{RP} - d_{RP}}\right] = |P_{RP}| \frac{d_{CP} - E[d_P]}{d'_{RP} - d_{RP}}. \qquad (8)$$

Thus, if the value obtained from Eq. (8) is smaller, the PDHT detection probability is larger. The derived equation indicates that the detection probability is larger (or smaller) if

1. The total delay added to the RP is larger (or smaller).

2. The number of gates in the RP is smaller (or larger).

3. The difference between the critical delay and average delay of paths (i.e., $E[d_P]$) is smaller (or larger).

In the following, we describe the relations between the above conditions and hardware algorithms for parallel multiplication. In general, a multiplier consists of three parts: a partial product generator, partial product accumulator (PPA), and final stage adder (FSA). In this study, we focus on the PPA and FSA because there are various algorithms for them and the latency and circuit area of a multiplier heavily depend on the algorithms. Here, we consider two typical types of PPAs: Wallace trees [11] and redundant binary addition (RBA) trees [12]. Wallace trees are

---

a uniformly distributed delay. However, it is too difficult to analyze the effects of such non-uniformly distributed delays.

among the fastest PPAs and are the optimal trees of three-input, two-output carry-save adders (CSAs) in terms of gate depth (i.e., delay). RBA trees are binary trees of four-input, two-output redundant CSAs based on a redundant binary representation. In an RBA tree, each digit of an integer is represented redundantly using 0, 1, and -1, which makes it possible to construct a PPA by using a symmetrical tree of four-input, two-output CSAs. Thus, Wallace trees have short delays while RBA trees have high efficiencies in terms of area-delay.

The path delay difference in a Wallace tree is large due to the asymmetrical tree structure. Thus, it would be difficult to detect PDHTs in multipliers with Wallace trees according to the above conditions. On the other hand, the path delay difference in an RBA tree is small owing to the symmetrical structure of binary trees, which implies that multipliers with RBA trees are more resistant to PDHT insertion than those with Wallace trees.

As FSAs, we focus on three typical carry-propagation adders (CPAs): ripple-carry adders (RCAs), block carry-lookahead adders (BCLAs) [13], and KoggeStone adders (KSAs) [14]. RCAs, BCLAs, and KSAs are among the optimal CPAs for circuit area, area-delay efficiency, and delay, respectively. RCAs are the simplest and most compact two-input CPAs. An n-bit RCA consists of n full adders, and the i-th ($0 \leq i \leq n-1$) full adder computes the i-th sum bit s_i and $(i+1)$-th carry bit c_{i+1} from the i-th input bit and i-th carry bit c_i. The critical delays of RCAs are the largest among the common CPAs due to the long carry propagation path. BCLAs have smaller critical delays than RCAs due to the use of the carry-lookahead technique. A BCLA consists of several small RCAs and a carry-lookahead unit. It generates some carry bits directly from the inputs using the carry-lookahead unit to make its critical delay smaller than that of an RCA. The main drawback of BCLAs is that their carry-lookahead units require gates with large fan-in and fan-out, which diminish the circuit performance. KSAs are the fastest CPAs based on parallel prefix operations, which define how to implement each carry bit generation block. KSAs perform addition with minimal gate depth (i.e., delay) of $O(\log(n))$ and only gate with a fan-out of two. Thus, KSAs usually achieve the smallest delays at the expense of circuit area. Since the lengths of most paths in RCAs are short, the average delay of an RCA is small relative to the critical delay. On the other hand, since BCLAs and KSAs calculate carries in parallel, many paths would be related to carry propagation. Therefore, KSAs and BCLAs have more long paths (whose delays are close to the critical delay) than RCAs. Thus, the differences between the critical and average delays for BCLAs and KSAs are small, and, consequently, PDHTs in multipliers with BCLAs and KSAs should be more easily detected than those in multipliers with RCAs according to the above conditions.

4 Experiments

Our argument in Section 3 was validated by experimentally inserting PDHTs into the six above-mentioned types of multipliers with 16-, 24-, and 32-bit operands combining two PPAs and three FSAs. In total, we evaluated 18 multipliers in the experiment. Note that such multipliers are frequently used for public key cryptography with more than 100-bit multiplication.

4.1 Evaluation of PDHT detection probability

To evaluate the probability of PDHT detection during Monte Carlo testing, we applied 10^7 random input vectors to PDHT-inserted multipliers and then counted the number of setup time violations by performing gate-level timing simulations. In addition, to analyze the dependencies of the setup time violations on the switching probabilities and gate counts of the delay-added paths, we randomly selected a path in each multiplier, added a delay to the selected path like a PDHT, and applied 10^6 random input vectors to the multiplier. We uniformly added a unit delay to each gate along the randomly selected path such that the resulting path delay was from 1.2 to 2.0 times the critical delay. For each multiplier, we repeated the above evaluation process 1,000 times.

Figure 3 presents the histograms of the numbers of such multipliers with randomly selected delay-added paths, where the horizontal axes indicate the detection probabilities of 10^7 random inputs, and the red, green, blue, cyan and purple bars denote the detection probabilities where the total path delays are from 1.2, 1.4, 1.6, 1.8 and 2.0 times the critical delay, respectively. Figures 4, 5 and 6 show the PDHT detection probabilities for all 18 multipliers, where the values of zero indicate that the corresponding RPs could not be detected in our experiment. From Figs 3 – 6, we can confirm that the average detection probabilities of the RCAs are smaller than those of the BCLAs and KSAs for any operand length and that all of the average detection probabilities tend to be smaller if the operand length is larger.

We discuss these results in detail in the following subsections.

4.2 Evaluation of path delays

We then evaluated the average path delay in each multiplier. As described in Section 3, Eq. (8) is essential for evaluating PDHT detectability (i.e., the number of setup time violations), which is closely related to the number of gates switched along the RP to cause setup time violation. Table 1 shows the critical delay, average delay of a randomly selected path, and difference between them for each multiplier. Table 1 reveals that the critical and average delays of the multipliers with Wallace trees are

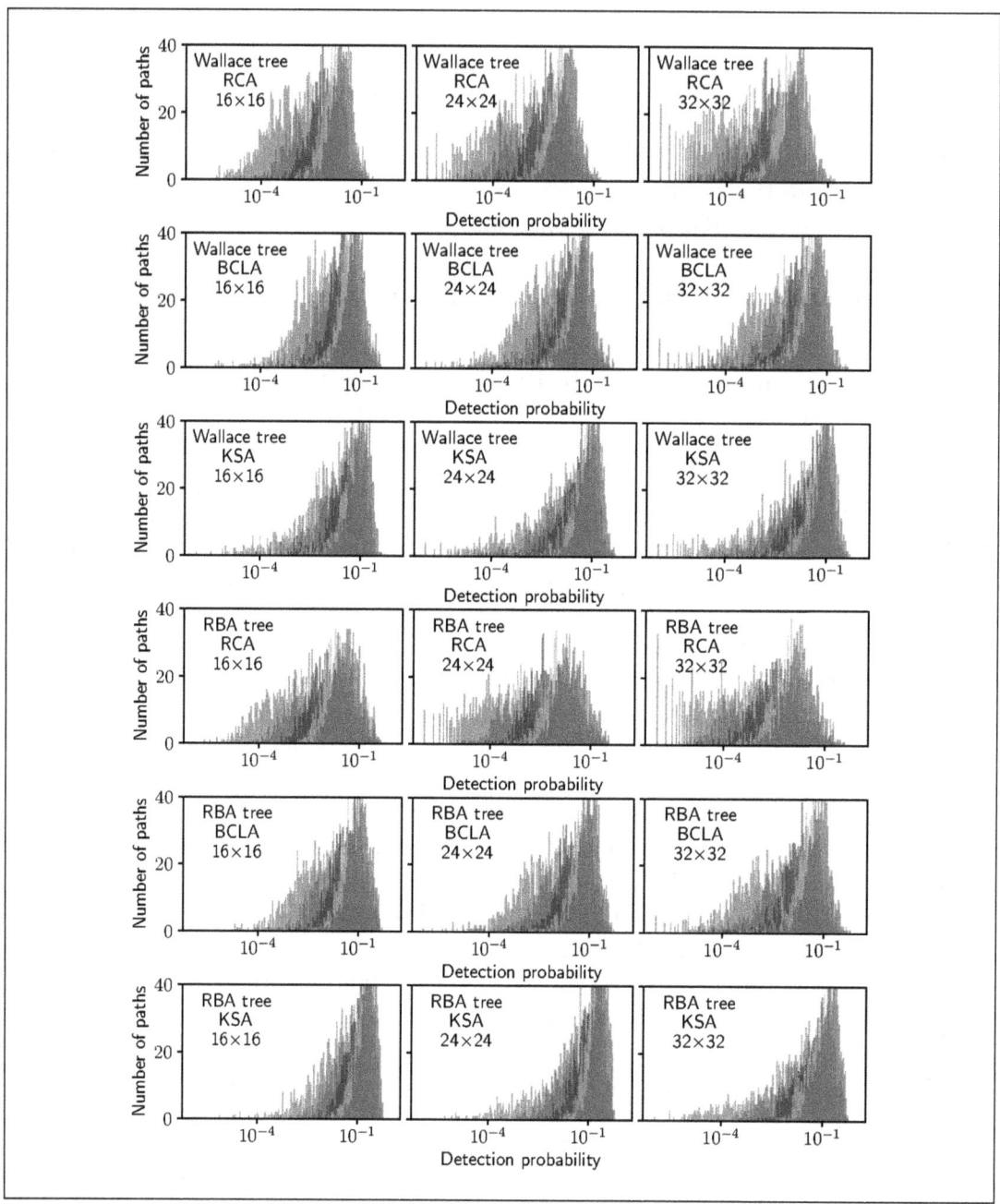

Figure 3: Detection probabilities of 1,000 randomly chosen paths.

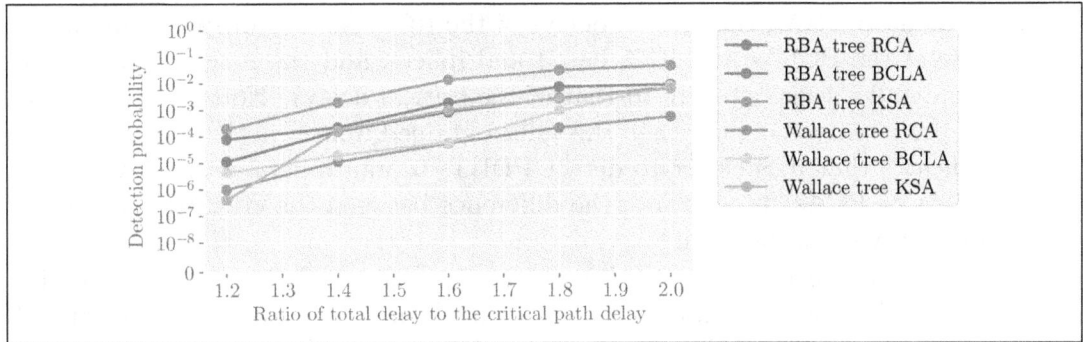

Figure 4: Detection probabilities of RPs inserted into 16-bit multipliers.

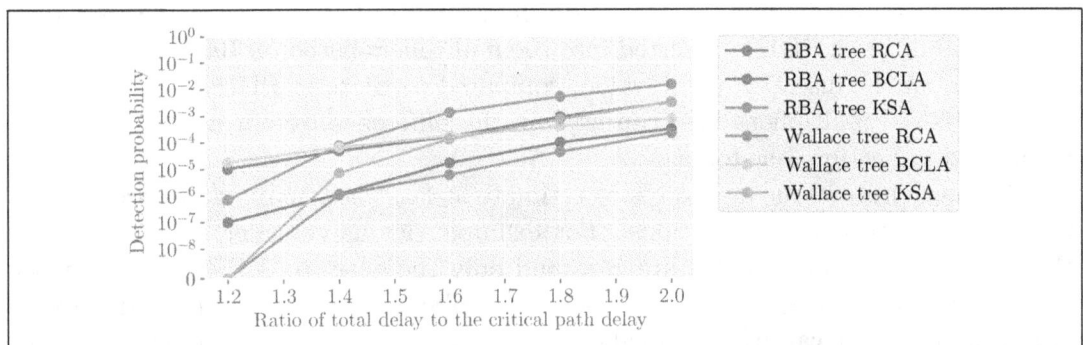

Figure 5: Detection probabilities of RPs inserted into 24-bit multipliers.

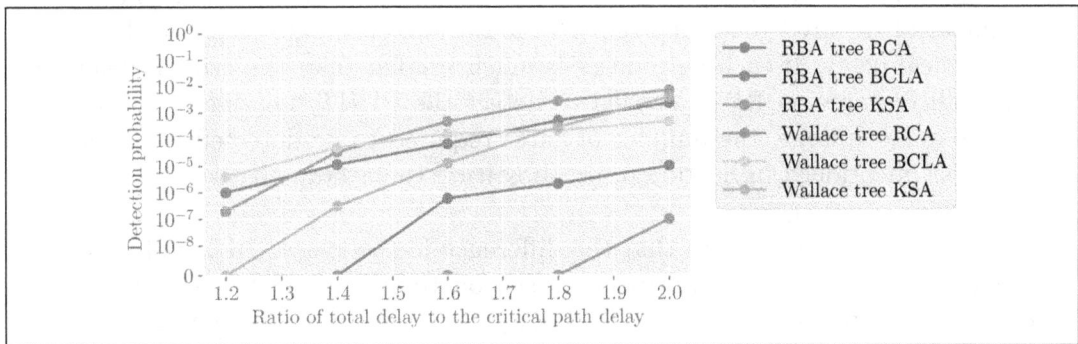

Figure 6: Detection probabilities of RPs inserted into 32-bit multipliers.

smaller than those of the multipliers with RBA trees, although both logic depths are given as $O(\log(n))$, where n is the number of input bits of the multiplier. This tendency occurred because extra logic for converting redundant binary into common binaries exists in multipliers with RBAs due to the redundant binary representation.

Regarding the FSAs, the critical delays of the RCA-based multipliers are larger than those of the BCLA- and KSA-based multipliers, and the magnitude relation of the average delays is opposite to that of the critical delays. Note again that the detection probability is higher when the value obtained from Eq. (8) is smaller. The result indicates that it is easier to detect PDHTs in multipliers with KSAs rather than BCLAs or RCAs. In contrast, the difference between the critical and average delays in a PPA would be trivial.

Based on the above results, we discuss the detection probabilities of randomly-inserted PDHTs where the insertion delays vary from 1.2 to 2.0 times the critical path delay in Fig. 3. From the figure, we can first confirm that the detection probabilities are lower when the insertion delay is smaller. This corresponds to the first condition described in Section 3. In addition, the result indicates that the detection probabilities of PDHTs inserted into the multipliers based on RCA vary greatly depending on the insertion delay value, compared to those of PDHTs inserted into the KSA based multipliers. We can explain the difference by our analysis method mentioned in Section 3 as follows.

The basic idea of our method is to estimate at least how many gates are needed to cause the setup time violations. For example, let us consider the case where the original delays of all gates are zero and only the gates on RP have the delays inserted by introducing the PDHT. The half number of gates on RP is needed to be switched in order to cause an error when the insertion delay is twice the critical delay. However, since the all gates on the circuit have delays in practice, the number of gates required to cause an error would be smaller than the half number. In addition, the number of required gates depends on the ratio of each gate delay on the circuit to the critical delay. If each gate delay is much smaller than the critical delay, the almost half of gates on RP is needed to detect the PDHT. On the other hand, if each gate delay is large, the number of gates required to be detected is smaller. In our analysis mentioned in Section 3, we substitute the average path delays for each gate delays.

From the table, we confirm that the difference in the average delay of randomly selected paths between the multipliers based on RCA and KSA is small. On the other hand, the difference of the critical delays between them is large. The number of gates needed to cause an error is larger in the case of RCA–based multipliers in comparison with KSA–based ones. This effect would be larger when the insertion delay is smaller. In our experiments, at the smallest insertion delay (i.e., 1.2 times the critical path delay), an error must cause when 83% of gates were activated. In the case of KSA–based multiplier, it is highly likely that the number of gates required to cause an error is actually smaller than the above number, but in the case of RCA, it is considered that almost the same number is required. As mentioned above, the

PPA	FSA	bit length	d_{CP} [ns]	$E[d_P]$ [ns]	Diff.
RBA tree	RCA	16	0.663	0.127	0.536
		24	0.975	0.144	0.830
		32	1.290	0.154	1.136
	BCLA	16	0.332	0.132	0.200
		24	0.347	0.149	0.198
		32	0.400	0.162	0.238
	KSA	16	0.257	0.142	0.115
		24	0.268	0.162	0.106
		32	0.310	0.171	0.139
Wallace tree	RCA	16	0.58	0.100	0.480
		24	0.885	0.116	0.769
		32	1.190	0.129	1.061
	BCLA	16	0.246	0.104	0.142
		24	0.290	0.123	0.167
		32	0.319	0.136	0.183
	KSA	16	0.197	0.111	0.085
		24	0.224	0.131	0.093
		32	0.247	0.142	0.105

Table 1: Critical and average delays of randomly chosen paths of various multipliers

number fo gates to cause an error is thought to be inversely proportional to the ratio of the insertion delay to the critical delay. In fact, the ratio appears in the denominator of Eq. 8 when we divide its numerator and denominator by the critical delay. From the above reasons, it cecomes more difficult to detect RBA rather than KSA and the detection becomes harder as the insertion delay decreases.

4.3 Evaluation of switching probability

The controllability of signals along the path significantly influences the detection probability because a random vector can easily sensitize a path consisting of gates with high switching probabilities. Figure 7 shows the histograms of the gate switching probabilities for 1,000 randomly selected paths in each multiplier, where the horizontal axes indicate the logarithmic mean switching probabilities. Table 2 also lists the corresponding average switching probabilities of the gates along the RPs.

Figure 7 confirms that the logarithmic means differ considerably depending on the PPA algorithm. The multipliers with Wallace trees have smaller logarithmic

PPA	FSA	16-bit	24-bit	32-bit
RBA tree	RCA	0.1589	0.1505	0.1523
RBA tree	BCLA	0.1099	0.0641	0.0455
RBA tree	KSA	0.1013	0.0459	0.0234
Wallace tree	RCA	0.0378	0.0317	0.0322
Wallace tree	BCLA	0.0640	0.0600	0.0517
Wallace tree	KSA	0.0641	0.0449	0.0374

Table 2: switching probabilities of RPs

mean switching probabilities than those with RBA trees. In other words, there are many paths with high activation probabilities in RBA trees, basically because Wallace trees have asymmetric structures including many gates with switching probabilities close to 0 or 1 while RBA trees have symmetric binary tree structures. The critical delays of the above FSAs and PPAs are given by $O(\log(n))$, except for the RCAs. Since the logic depths of the paths in such FSAs and PPAs increase gradually with increasing operand length, the switching probability of each path does not strongly depend on the operand length. Similarly, as mentioned in Section 3, many paths in an RCA have far smaller logic depths than the critical path, which indicates that the switching probability also does not strongly depend on the operand length for an RCA.

4.4 Evaluation of our method

To conduct a theoretic validation of our argument, we performed multiple regression analysis between the detection probabilities of 1,000 paths randomly selected from each multiplier and predictor variables: (i) the controllabilities, (ii) the numbers of gates along randomly selected paths, and (iii) M, given by $(d_{CP} - E[d_P])/(d'_{RP} - d_{RP})$. Note that each variable is normalized to unit variance in order to make the influence of the variable on the multiple regression model meaningful and understandable. In this evaluation, we focus on 32-bit multipliers, which are more frequently used in practical cryptographic HW than 16- and 24-bit ones. Table 3 shows the multiple regression analysis coefficients and t-stats. The R-squared value obtained from the multiple regression was 0.699. Table 3 confirms that |t-stat| is large for M and the switching probability, which indicates their significant influence on the model. Thus, M and the switching probability are good variables for explaining PDHT detectability.

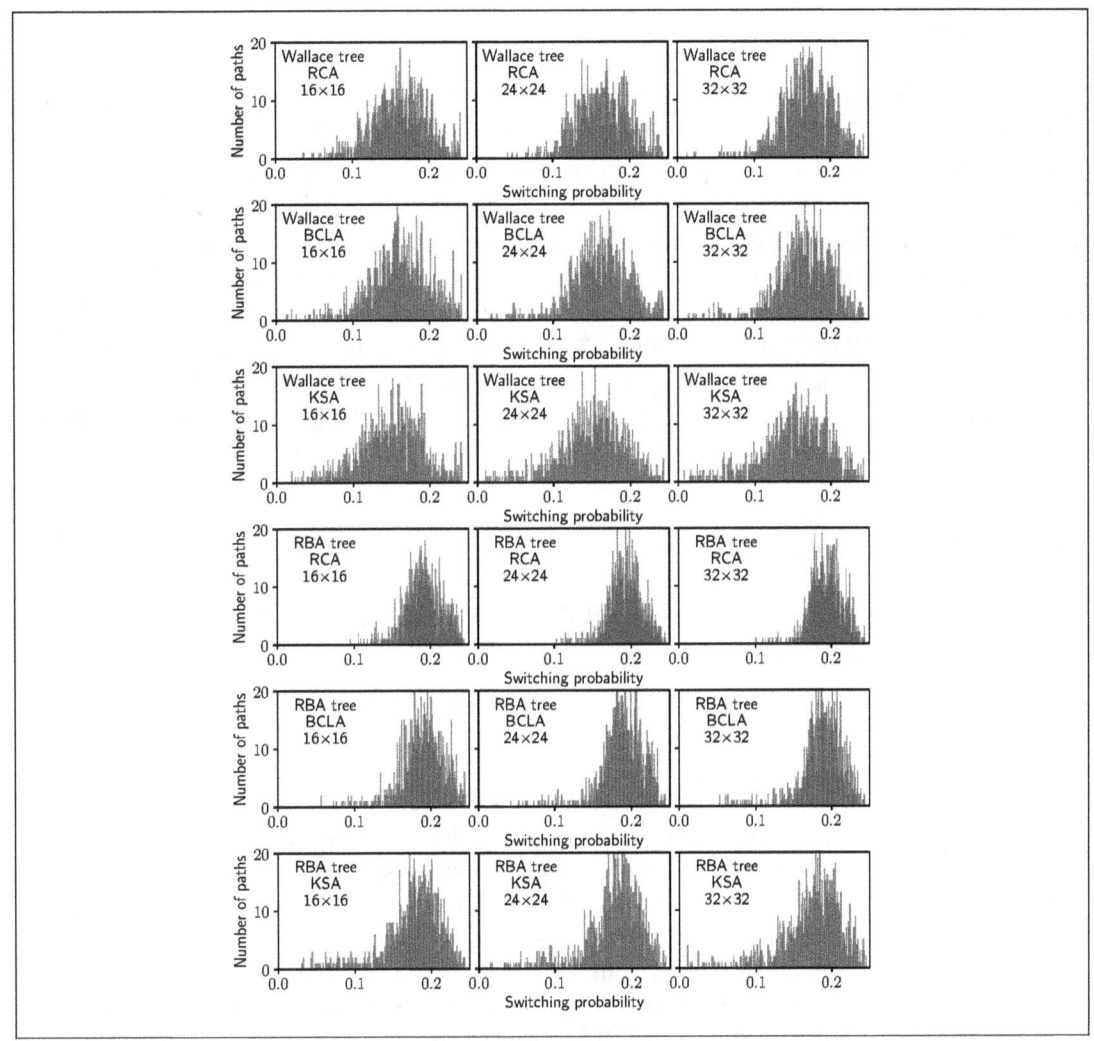

Figure 7: Switching probabilities of 1,000 randomly chosen paths.

	M	Number of gates	Switching probability
coefficient	-530.1	34.88	163.1
t-stat	-255.6	17.21	59.86

Table 3: Results of multiple regression analysis

5 Conclusion

This paper presented analyses of parallel multiplication hardware algorithms from the viewpoints of RP characteristics and PDHT detectability. We discussed the theoretical aspects of RPs in multipliers and their relations to PDHT detectability. Our argument was validated through the experimental insertion of PDHTs into various multipliers. In addition, we confirmed the effectiveness of our method through the experimental insertion with different amounts of inserted delays. The results of multiple regression analysis confirmed that the proposed evaluation method primarily explains PDHT detectability. The multiplier combined with an RBA tree and KSA yielded the highest detectability among the evaluated multipliers. The development of dedicated multiplication hardware algorithms that impede PDHT insertion remains a topic for future work.

Acknowledgment

This research has been supported by JSPS KAKENHI Grants No. 17H00729, No. 16K12436 and No. 16J05711.

References

[1] Shivam Bhasin, Jean-Luc Danger, Sylvain Guilley, Xuan Thuy Ngo, and Laurent Sauvage. Hardware trojan horses in cryptographic ip cores. In *Fault Diagnosis and Tolerance in Cryptography (FDTC), 2013 Workshop on*, pages 15–29. IEEE, 2013.

[2] Eli Biham and Adi Shamir. Differential fault analysis of secret key cryptosystems. *Advances in Cryptology—CRYPTO*, pages 513–525, 1997.

[3] Georg T Becker, Francesco Regazzoni, Christof Paar, and Wayne P Burleson. Stealthy dopant-level hardware trojans. In *International Workshop on Cryptographic Hardware and Embedded Systems*, pages 197–214. Springer, 2013.

[4] Samaneh Ghandali, Georg T Becker, Daniel Holcomb, and Christof Paar. A design methodology for stealthy parametric trojans and its application to bug attacks. In *International Conference on Cryptographic Hardware and Embedded Systems*, pages 625–647. Springer, 2016.

[5] Ronald L Rivest, Adi Shamir, and Leonard Adleman. A method for obtaining digital signatures and public-key cryptosystems. *Communications of the ACM*, 21(2):120–126, 1978.

[6] Neal Koblitz. Elliptic curve cryptosystems. *Mathematics of computation*, 48(177):203–209, 1987.

[7] Eli Biham, Yaniv Carmeli, and Adi Shamir. Bug attacks. In *Annual International Cryptology Conference*, pages 221–240. Springer, 2008.

[8] Koren Israel. Computer arithmetic algorithms. *AK Peters, Ltd*, 2002.

[9] Gordon L Smith. Model for delay faults based upon paths. In *ITC*, pages 342–351, 1985.

[10] Sunil K Jain and Vishwani D Agrawal. Stafan: An alternative to fault simulation. In *Proceedings of the 21st Design Automation Conference*, pages 18–23. IEEE Press, 1984.

[11] Christopher S Wallace. A suggestion for a fast multiplier. *IEEE Transactions on Electronic Computers*, (1):14–17, 1964.

[12] Naofumi Takagi, Hiroto Yasuura, and Shuzo Yajima. High-speed vlsi multiplication algorithm with a redundant binary addition tree. *IEEE Transactions on Computers*, (9):789–796, 1985.

[13] Amos R Omondi. Computer arithmetic systems algorithms, architecture and implementation. 1994.

[14] Peter M Kogge and Harold S Stone. A parallel algorithm for the efficient solution of a general class of recurrence equations. *IEEE Transactions on Computers*, 100(8):786–793, 1973.

A NOISE-SHAPING ANALOG-TO-DIGITAL CONVERTER USING A $\Delta\Sigma$ MODULATOR FEEDFORWARD NETWORK

TAKAO WAHO[*]
Graduate School of Science and Technology
Sophia University
7-1 Kioicho, Chiyoda, Tokyo 102-8554, Japan
t-waho@sophia.ac.jp

Abstract

A noise-shaping analog-to-digital converter (ADC) using a $\Delta\Sigma$ modulator network is proposed, and signal-level simulations are carried out as a proof of concept. The present architecture is based on a feedforward artificial neural network, where an N-bit digital output is generated through N channels containing one $\Delta\Sigma$ modulator per channel. A moving average taken from each $\Delta\Sigma$ modulator is optimized to obtain a multi-level feedforward signal. Simulation results show proper noise-shaping characteristics for both the first-order and second-order $\Delta\Sigma$ modulators. The effective number of bits (ENOB) increases as the number of channels increases up to six. A non-binary conversion scheme suggests a further advance in the ENOB. Finally, the present ADC is compared with conventional multi-bit $\Delta\Sigma$ modulators.

1 Introduction

As an interface between the real analog world and digital systems, analog-to-digital converters (ADCs) are indispensable in wireline/wireless communications, consumer electronics, and sensor networks [1, 2]. They are also expected to play a key role even in the future Internet of Things (IoT). These applications strongly require low power consumption as well as a bit resolution of 8 bits or higher. A good balance between these performances and the operation speed is also an important design challenge. Various architectures have been investigated to meet these conditions, including successive approximation, pipelined, and $\Delta\Sigma$ architectures [3]. Another

---

[*]This work was supported by JSPS KAKENHI Grant Number JP15K06030.

interesting architecture, which we have focused on, is based on the artificial neural network (ANN) first proposed in 1986 [4].

In the ANN-based ADC architecture, the input analog signal is applied to several channels in a neural network, and neurons at the end of the channels generate the output bits. Multilevel or analog signals flow between the channels to perform A/D conversion. This makes the circuit configuration compact. The learning capability inherent to the ANN is also attractive to achieve self-calibration, which is essential for improving the ADC resolution [5]. However, ANN-based ADCs suffered from the local minimum problem, which could result in unacceptably large conversion errors [6].

To solve the problem, simple comparators conventionally used as neurons were replaced with $\Delta\Sigma$ modulators [7]. Signal-level simulation showed that the bit resolution was increased to more than 15 bits by increasing the oversampling ratio (OSR) to around 1000. Noise shaping characteristics, being unique to the $\Delta\Sigma$ modulation, were also observed. An alternative method was also presented to further increase the conversion resolution [8]: *i.e.*, Increasing the number of channels in the ANN network instead of increasing the OSR. Taking the moving average of a 1-bit $\Delta\Sigma$ modulator output was proposed to generate multi-level signals, which were then used as intermediate signals connecting the channels.

The purpose of this paper is to further discuss the alternative method [8] by presenting detailed simulation results, and to confirm its effectiveness as a "hybrid" approach to high-performance ADCs. In Section II, ANN-based ADC architectures are briefly reviewed, and the present approach is described. Then, signal-level simulation results are presented in Section III as a proof of concept, followed by discussion in Section IV.

2 ANN-based ADC Architectures

A Hopfield-type 4-channel ANN-based ADC is shown in Fig. 1(a) [4]. The output of channels represents the digital output. D1 and D4 are the most significant bit (MSB) and the least significant bit (LSB), respectively. In the work of Tank and Hopfield [4], sub-ADCs were simply comparators. However, they suffered from the local minimum problem that caused a large quantization error. To solve it, an asymmetric network shown in Fig. 1(b) was proposed [9–11]. This network is equivalent to a feedforward one shown in Fig. 1(c), which is the basis of the present architecture.

An alternative method to suppress the local minimum effect is to replace the comparators with $\Delta\Sigma$ modulators [7]. Apparently random nature of $\Delta\Sigma$ modulator

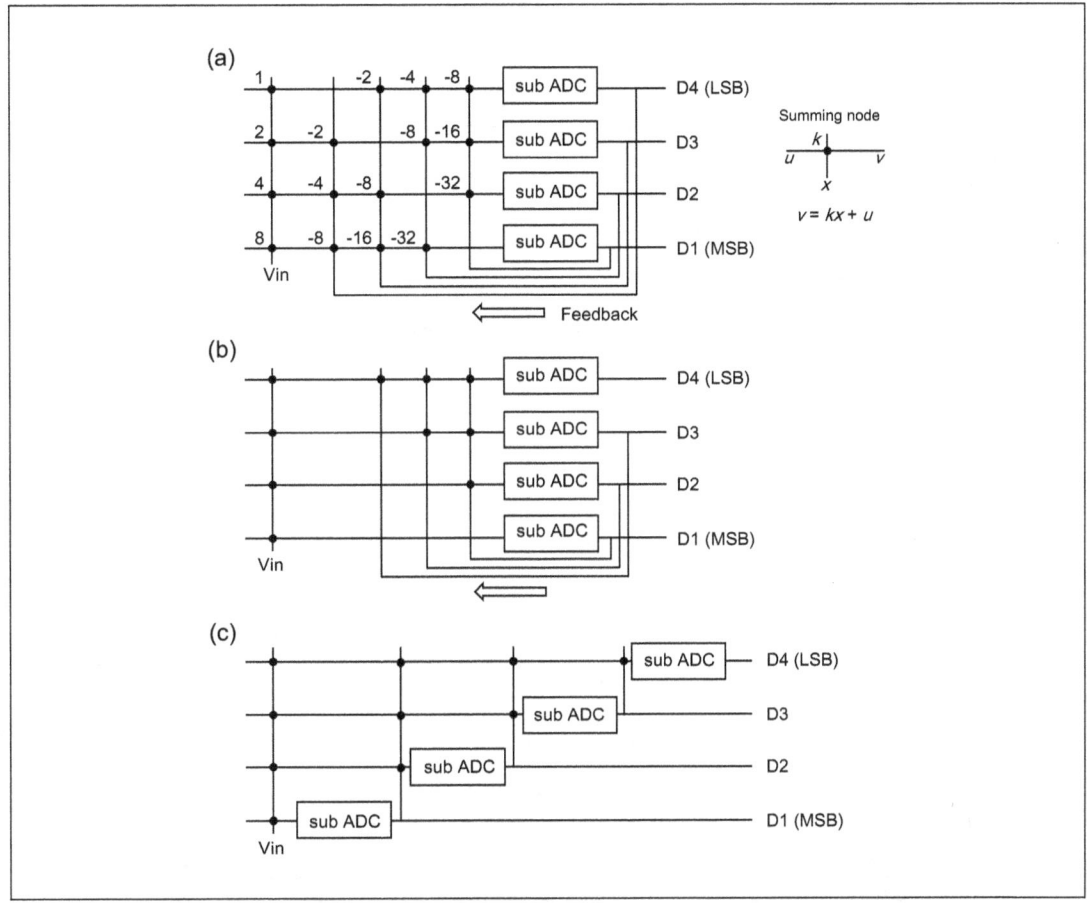

Figure 1: Block diagrams of 4-channel ADCs using Hopfield (a), asymmetric (b), and feedforward (c) networks.

outputs effectively recovers the system from a local minimum. On the basis of this, a feedforward architecture using $\Delta\Sigma$ modulators as neurons is investigated in this study. As an example, the present architecture consisting of four channels is shown in Fig. 2. This converts the analog input, x, into the digital output, y. DSM_1 to DSM_4 are $\Delta\Sigma$ modulators used as sub-ADCs. In the present study, a 1-bit first-order or second-order modulator shown in Fig. 3 was assumed. The constant α is the radix in the conversion, which is equal to 2 unless otherwise mentioned. Circled A's represent to take a moving average, which was introduced in Ref. [8]. As will be proved later, this was essential to improve the conversion resolution, because a moving average can effectively suppress high-frequency quantization noises due to

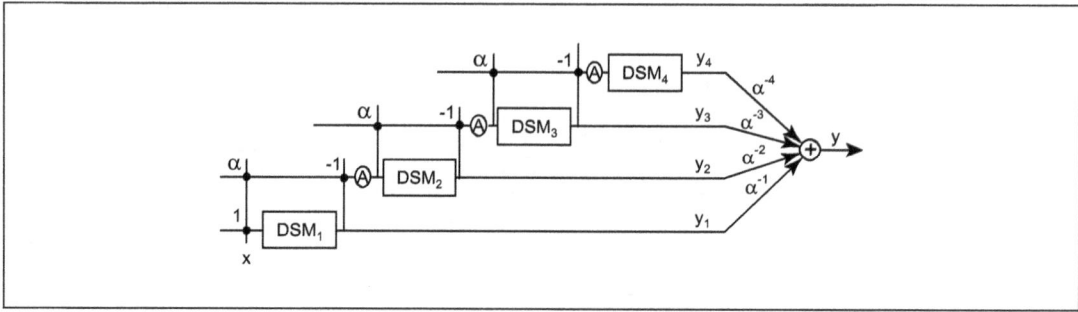

Figure 2: Proposed 4-channel ADC based on a feedforward network. Circled A's represent a moving average.

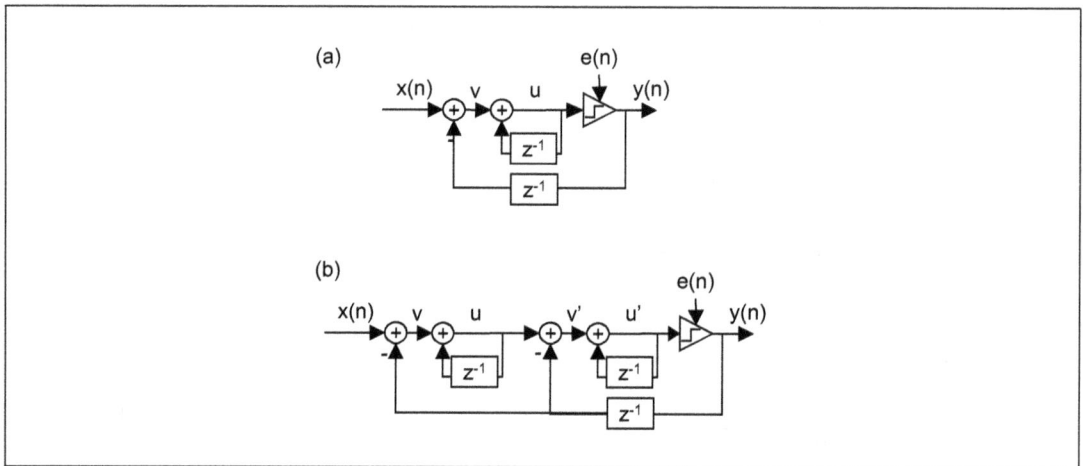

Figure 3: Block diagrams of first-order (a) and second-order $\Delta\Sigma$ modulators used in this study.

the noise shaping in $\Delta\Sigma$ modulation.

A 2-channel ADC with $\alpha = 2$ is depicted in Fig. 4. Here, the outputs of $\Delta\Sigma$ modulators are represented as

$$y_1(n) = x(n) + E_1(n) \tag{1}$$

and

$$y_2(n) = q(n) + E_2(n). \tag{2}$$

Here, $q(n)$ is a moving average of $p(n)$, which is written as

$$q(n) = \overline{p(n)} = \frac{1}{M} \sum_{i=n-M+1}^{n} p(i), \tag{3}$$

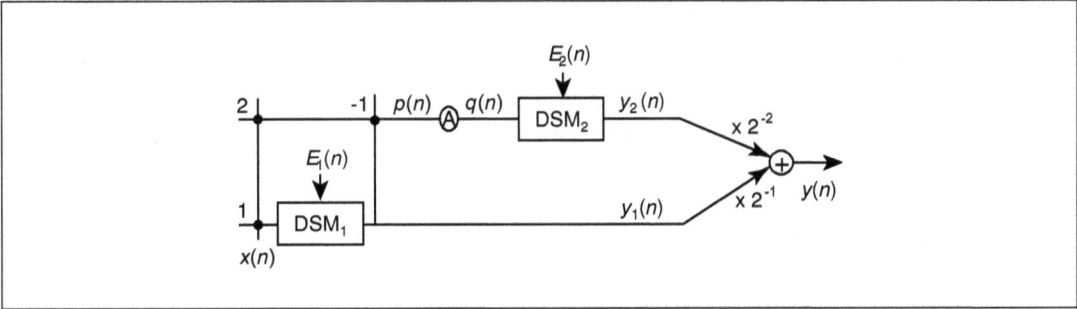

Figure 4: Proposed 2-channel ADC based on a feedforward network. Radix α in Fig. 2 is 2, resulting in a 4-level output.

where
$$p(i) = 2x(i) - y_1(i), \qquad (4)$$

and M is the number of samples taken for the moving average. $E_i(n)$ is the quantization error associated with the $\Delta\Sigma$ modulation. If a first-order modulator is used,
$$E_i(n) = e_i(n) - e_i(n-1), \qquad (5)$$

where $e_i(n)$ is the quantization error due to the quantizer ($e(n)$ shown in Fig. 3) in the modulator [12]. For a second-order $\Delta\Sigma$ modulator,
$$E_i(n) = (e_i(n) - e_i(n-1)) - (e_i(n-1) - e_i(n-2)). \qquad (6)$$

The input to DSM$_2$ $q(n)$ is an $(M+1)$-valued feedforward signal from the first channel to the second channel. The output $y(n)$ is obtained as
$$y(n) = y_1(n)2^{-1} + y_2(n)2^{-2}. \qquad (7)$$

In this 2-channel case, the output $y(n)$ is one of four values (-0.75, -0.25, +0.25, and +0.75) because $y_i(n) = \pm 1$. Similarly, for an N-channel case, $y(n)$ is a 2^N-valued output.

3 Simulation results

3.1 Output waveform and spectrum

The sinusoidal input and corresponding output waveforms obtained from the 2-channel structure shown in Fig. 4 are presented in Fig. 5. As shown in this figure,

$y_1(n)$ and $y_2(n)$ are normal $\Delta\Sigma$ modulator outputs for the input signals $x(n)$ and $q(n)$ respectively, $q(n)$ is the moving average of $p(n)(=2x(n)-y_1(n))$, and $y(n)$ is the 4-valued output as was described in Eq. 7. In the present study, $M=8$ was assumed unless otherwise mentioned. The operation of the present ADC is thus confirmed.

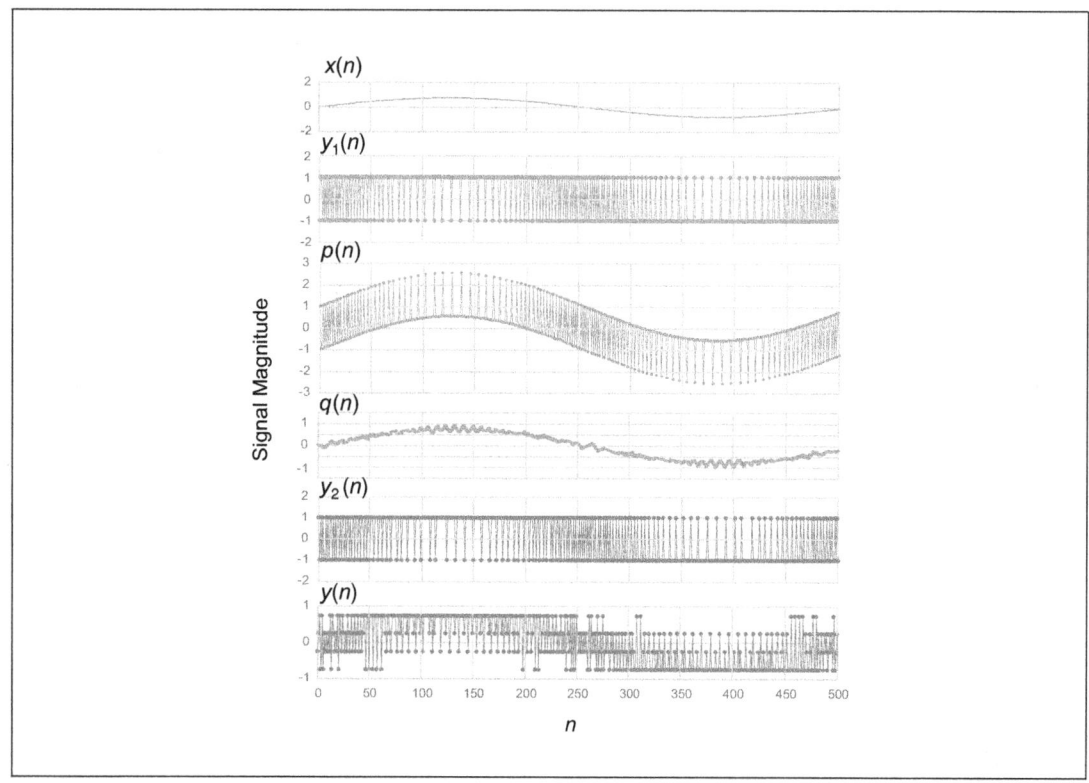

Figure 5: Output waveforms at several nodes in Fig. 4.

The input and output waveforms obtained from a present 4-channel ADC as shown in Fig. 2 with $\alpha=2$ is also plotted in Fig. 6. In this case, the output $y(n)$ is a 16-level signal (-0.9375, -0.8125, -0.6875, $\cdots$, +0.9375) as was explained above. In other words, the analog input is represented by the oversampled 16-level output. The output spectrum obtained by FFT analysis is shown in Fig. 7. Clear noise-shaping characteristics with a slope of about 20dB/dec indicate a proper first-order $\Delta\Sigma$ modulation. Note that simulation conditions used to obtain the time-domain and frequency-domain results were different from each other for the purpose of showing them clearly.

The sinusoidal input and corresponding output waveforms obtained from a 4-

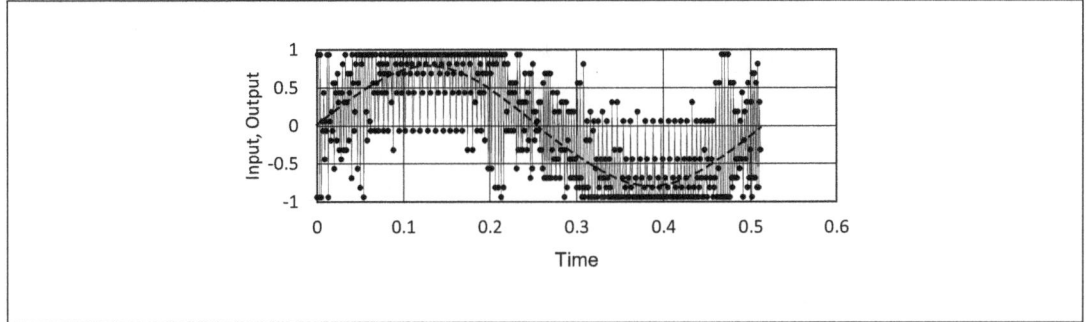

Figure 6: Input sinusoidal (dashed line) and 16-level output waveforms obtained from a present 4-channel ADC using four first-order $\Delta\Sigma$ modulators.

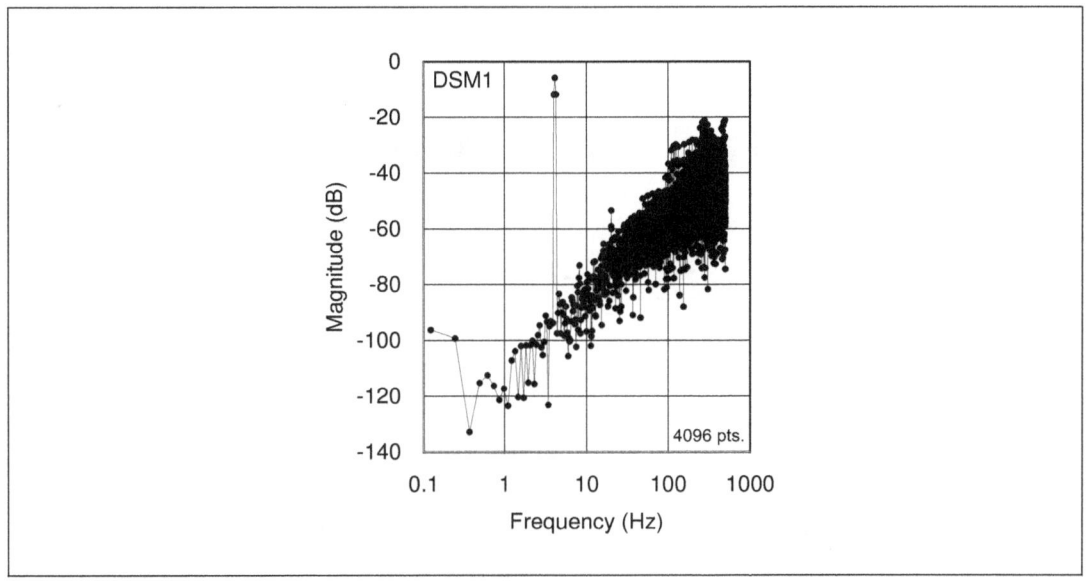

Figure 7: Spectral density obtained from a present 4-channel ADC using first-order $\Delta\Sigma$ modulators.

channel ADC with second-order $\Delta\Sigma$ modulators are shown in Fig. 8. The output $y(n)$ is, again, a 16-level signal: -0.9375, -0.8125, -0.6875, $\cdots$, +0.9375. Although the difference between the output waveforms of Figs. 6 and 8 is not clear, the output spectrum shown in Fig. 9 reveals a clear difference in a noise-shaping slope; it is almost 40dB/dec, being twice as large as that in Fig. 7. This shows the second-order $\Delta\Sigma$ noise-shaping characteristics.

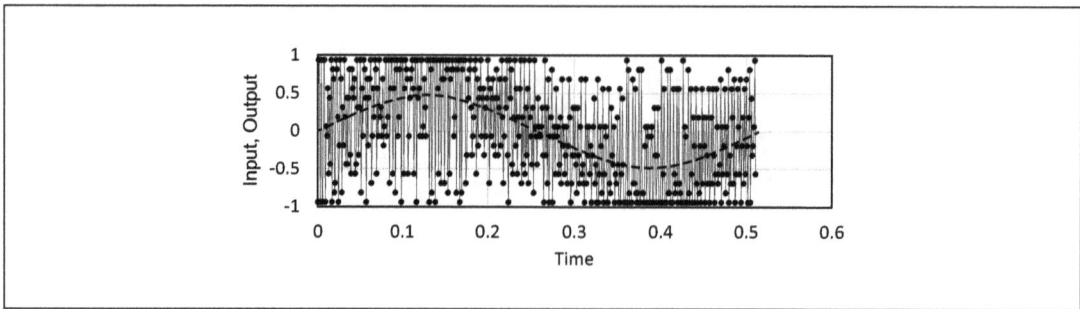

Figure 8: Input sinusoidal (dashed line) and 16-level output waveforms obtained from a present 4-channel ADC using second-order ΔΣ modulators.

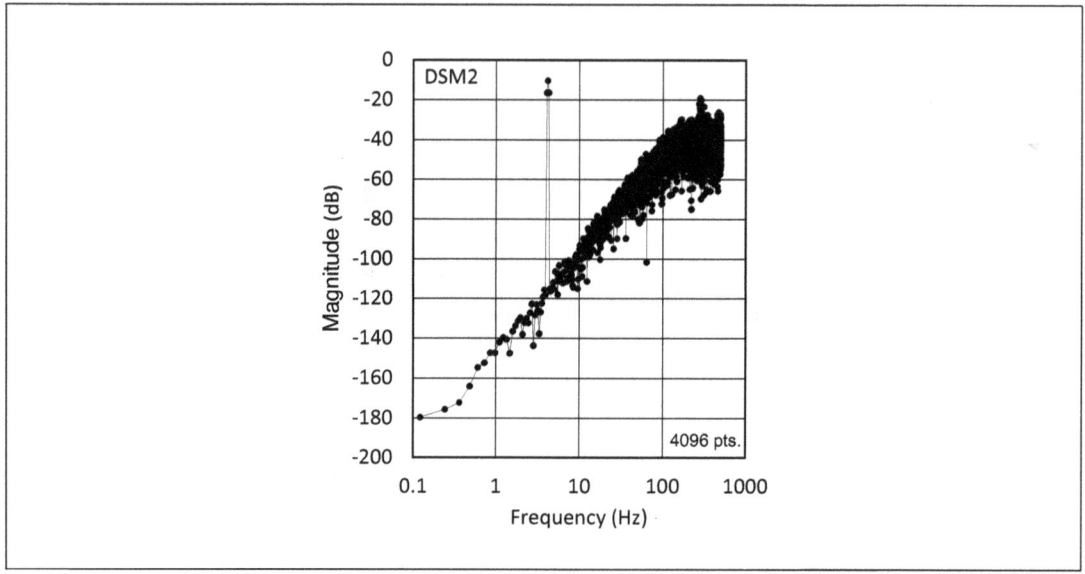

Figure 9: Spectral density obtained from a present 4-channel ADC using four second-order ΔΣ modulators.

3.2 Bit resolution

From the output spectrum shown in Figs. 7 and 9, the bit resolution of ADCs (more specifically, the effective number of bits (ENOB) [1]) can be calculated. The increase in ENOB indicates an improvement in the bit resolution. Fig. 10(a) shows ENOB values obtained from the present ADC with first-order ΔΣ modulators. Each plot corresponds to a single output spectrum shown in Fig. 7. Because the ENOB is equivalent to the signal-to-noise ratio, and because the quantization noise power is

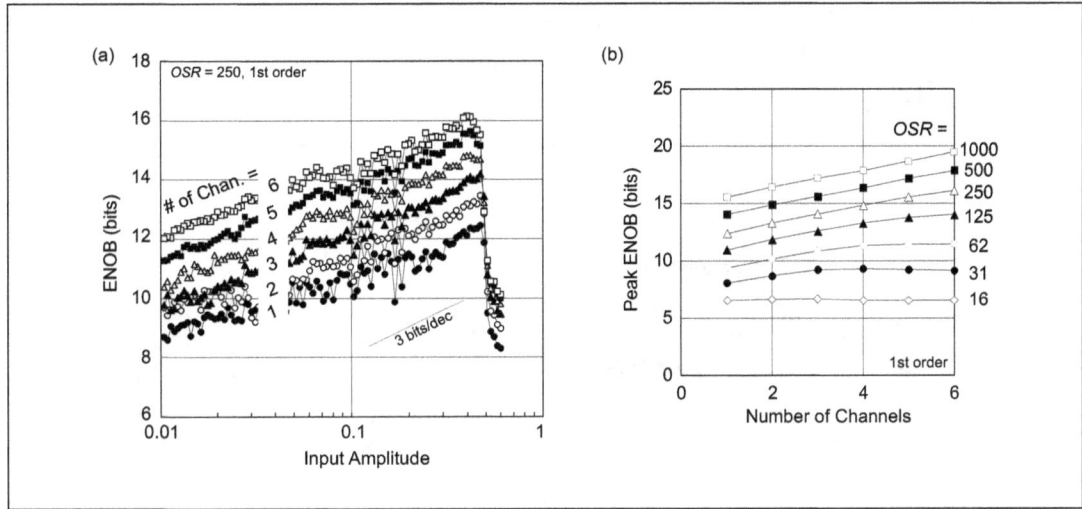

Figure 10: Effective number of bits (ENOB) as a function of the input amplitude obtained from present ADCs using first-order $\Delta\Sigma$ modulators (a) and peak ENOB as a function of the number of channels (b).

independent of the input amplitude, the ENOB improves as the input amplitude increases. The slope can be theoretically predicted to be 3 bits/dec, which agrees well with the simulation results. This figure also shows that the ENOB increases as the number of channels increases from one to six. The input full-scale was assumed to be ± 0.5. As the amplitude approaches the full-scale value, the ENOB saturates and drops abruptly. This abrupt decrease in ENOB results from the instability of $\Delta\Sigma$ modulators. In other words, if the input value exceeds the feedback signal in the $\Delta\Sigma$ loop, ± 0.5 in this case, the $\Delta\Sigma$ loop becomes unstable.

From Fig. 10(a), the peak ENOB was obtained as an average over the three largest ENOB values to avoid a fluctuation effect in the estimated ENOB. Fig. 10(b) shows the results obtained from ADCs with first-order $\Delta\Sigma$ modulators. When an oversampling ratio (OSR) is larger than 500, the peak ENOB increases by 1 bit for every increase in the number of channels. This shows that every channel in the present ADC provides 1-bit digital output, just as the same in the conventional neural-network ADCs. In contrast, however, the increase in the peak ENOB saturated for smaller OSR values. This will be discussed below.

Fig. 11 shows the ENOB and the peak ENOB obtained from the ADCs with second-order $\Delta\Sigma$ modulators. For $OSR > 500$, it is observed, again, that the peak ENOB increases by 1 bit for every increase in the number of channels. Fig. 11(a) shows that the input amplitude corresponding to the maximum ENOB is smaller for the second-order $\Delta\Sigma$ ADCs than for the first-order ones. This is because the second-order modulator tends to be unstable for smaller input values than the first-order modulator [12].

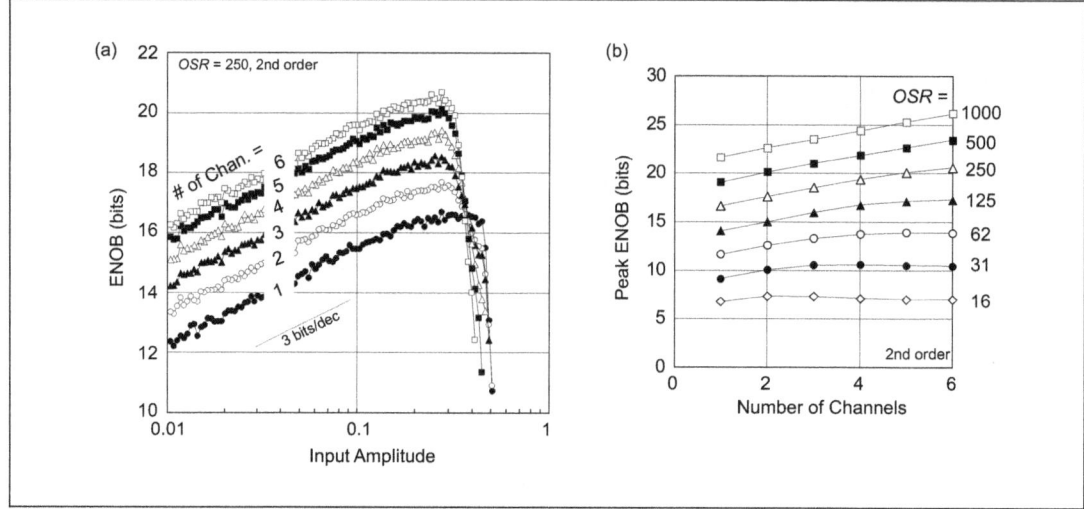

Figure 11: Effective number of bits (ENOB) as a function of the input amplitude obtained from present ADCs using second-order $\Delta\Sigma$ modulators (a) and peak ENOB as a function of the number of channels (b).

The peak ENOBs obtained for $OSR = 125$ are replotted in Fig. 12 as well as well as those without taking the moving averages. As was explained above, the peak ENOB with the moving averages increases as the number of channels increases. On the other hand, without the moving average, it was not enhanced even if the number of channels increased. In the second-order modulator case, the ENOB became even worse when the number of the channels increased. Therefore, it is proved that taking the moving average, as shown in Figs. 2 and 4, is indispensable to increase the ENOB, or to improve the bit resolution by effectively suppress high-frequency quantization noises due to $\Delta\Sigma$ modulation.

To clarify the effect of moving average, the peak ENOB was calculated for various sample numbers in the moving average. Fig. 13 shows the results. By taking the moving average, high-frequency quantization noise components is suppressed, which results in the increase in the peak ENOB. However, if the sample number in the moving average is too large, the signal can be blurred out and the peak ENOB decreases. Therefore, the optimum number of samples for the best peak ENOB exists, as shown in this figure. In this simulation, since the sampling period was kept constant, the optimum number for $OSR = 500$ is 4 times larger than that

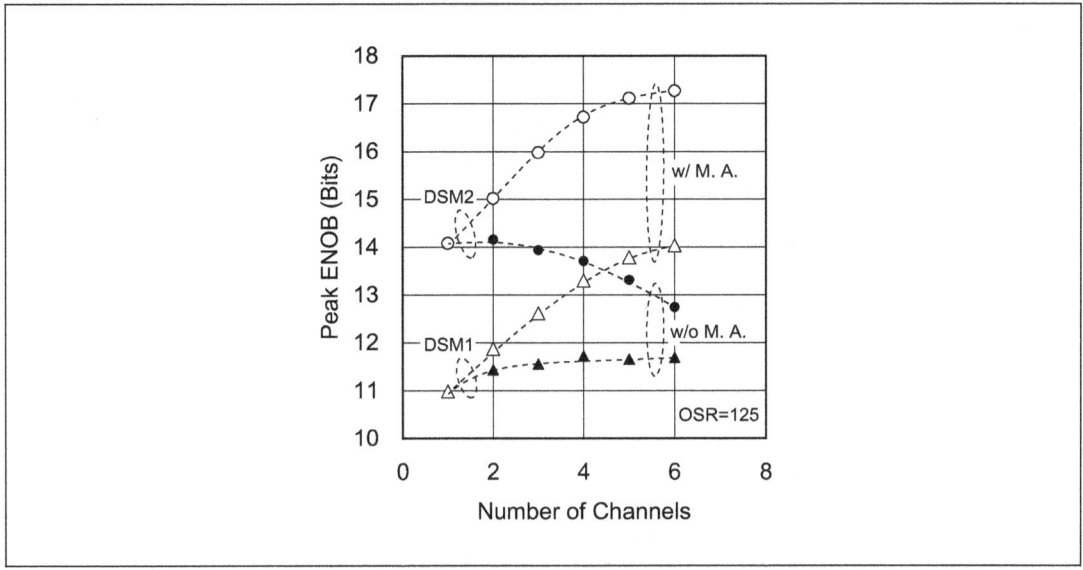

Figure 12: Peak ENOB as a function of the number of channels obtained from present ADCs using first-order (DSM1) and second-order (DSM2) $\Delta\Sigma$ modulators with and without moving averages (M. A.).

for $OSR = 125$. Also, it should be mentioned that the peak ENOBs for 4-channel and 6-channel ADCs are almost the same if $OSR = 125$ and $M/2 > 10$. This agrees with the saturations in the peak ENOB shown in Figs. 10(b) and 11(b). The moving average is a lowpass filter, the bandwidth of which is inversely proportional to the sample number in the average. Thus, as the ample number increases, the bandwidth decreases. It should be also noted that this lowpass filter is cascaded in the present ADC as shown in Fig. 2. Therefore, the more the number of channels, the narrower the bandwidth, which probably results in the saturation. In fact, for $OSR = 16$, since the input wavelength is only 4 times longer than the moving average period, only a small portion of the input signal is transmitted to higher channels, and therefore, virtually no improvement was obtained in the peak ENOB.

Results presented above were obtained by assuming $\alpha = 2$, or the binary conversion. Recently, non-binary schemes are sometimes effectively used in ADCs [13]. By introducing redundancy, the requirement on the mismatch in device characteristics can be relaxed, resulting in the better ENOB. Fig. 14 shows the results obtained by changing the value of α. It is found that $\alpha = 2$ is a reasonable choice both for first- and second-order $\Delta\Sigma$ modulators. Slightly better ENOBs are observed in the second-order case, which might be related to the potential instability. Reducing the

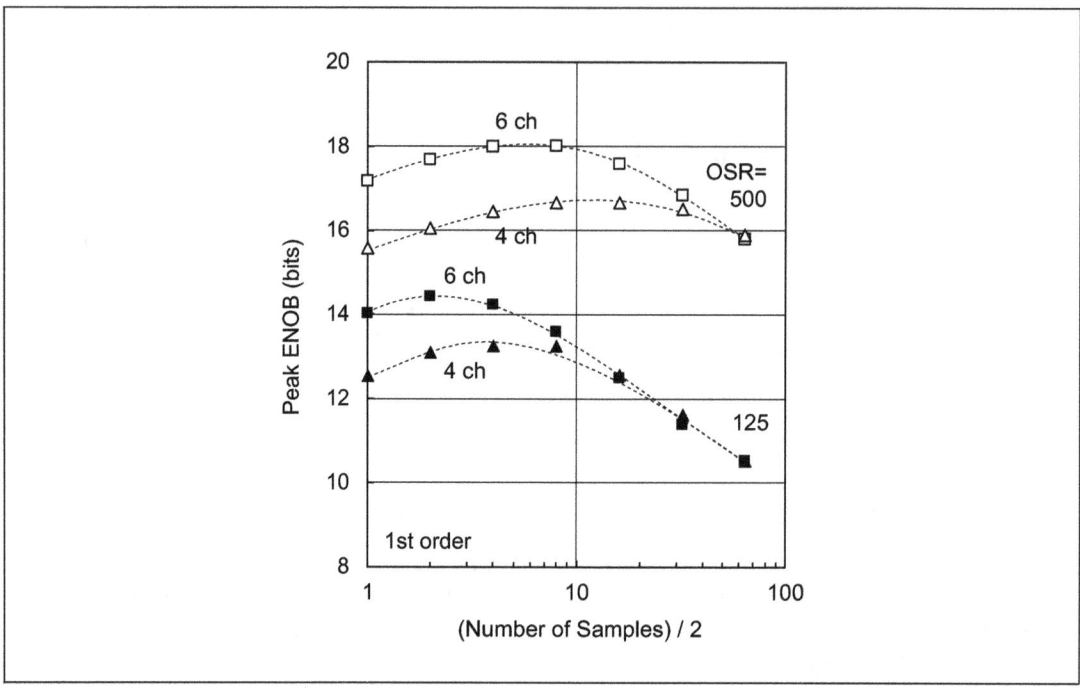

Figure 13: Peak ENOB as a function of the number of samples in moving averages. Note that the horizontal axis is $M/2$.

radix can effectively improve the stability by introducing a redundancy. This figure suggests that high accuracy in α is not required because the ENOB remains almost constant even if α varies from 1.8 by $\pm 10\%$, for example. However, α at the output summing node should be matched with α at the input node.

4 Discussion

Conventional $\Delta\Sigma$ modulators with architecture similar to the present ADC shown in Fig. 2 are known as cascaded multi-stage $\Delta\Sigma$ modulators, or MASH [14], and multibit $\Delta\Sigma$ modulators [12]. In the MASH architecture, the difference between the output and input of the comparator in the $\Delta\Sigma$ modulator is used as the feedback signal. In contrast, in the present structure, the difference between the terminals of the $\Delta\Sigma$ modulator itself, for instance DSM_1, is used as the feedforward signal. While higher-order operation is achieved in the MASH architecture, the multi-bit operation, as shown in Figs. 10 and 11, is achieved in the present architecture. These two operation modes can be used complementarily, or combined into a single

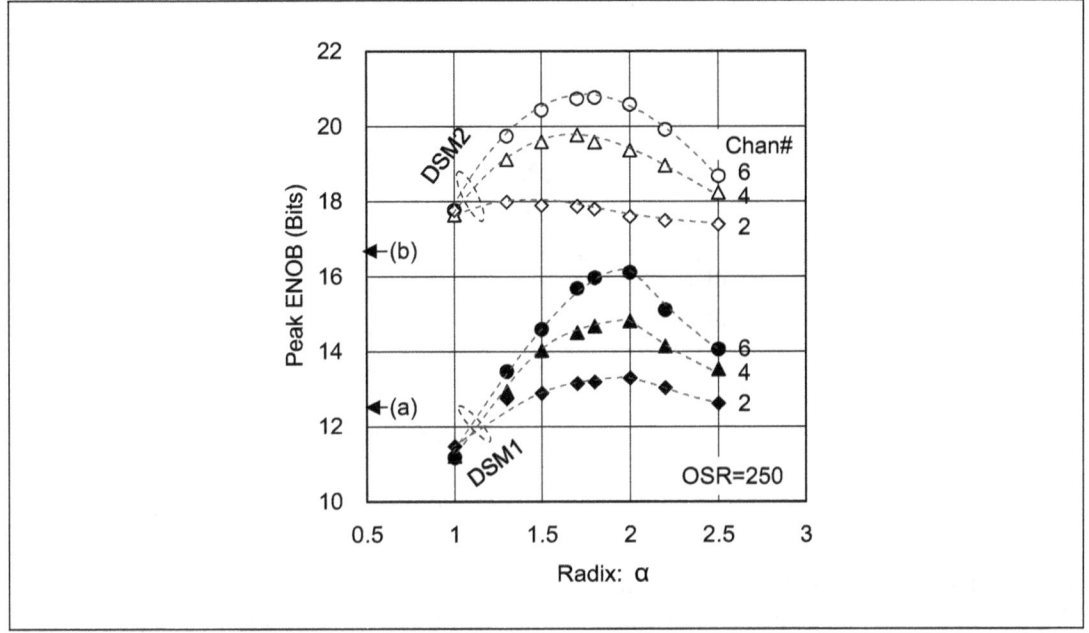

Figure 14: Peak ENOB as a function of the radix α obtained from present ADCs using first-order (DSM1) and second-order (DSM2) $\Delta\Sigma$ modulators with 2, 4, and 6 channels. For comparison, peak ENOB for conventional first-order and second-order $\Delta\Sigma$ modulators are shown by arrows (a) and (b), respectively.

ADC in the future.

It is interesting to compare the present ADC with conventional multibit $\Delta\Sigma$ modulators. Fig. 15 compares the peak ENOB of the present ADCs with that of conventional multibit $\Delta\Sigma$ modulators. For the first-order case, the present and conventional configurations show the peak ENOB comparable to each other. For the second-order case, the peak ENOB of conventional multibit modulators is better than that of the present ADC for the number of bits larger than 3. The reason for this is related to the stability of the $\Delta\Sigma$ negative feedback loop. As the number of bits increases, the quantization noise decreases, which results in a better stability. This allows the multibit modulator to operate with a larger input amplitude, resulting a higher peak ENOB. In contrast, the present ADC uses 1-bit modulators in each channel, so even if the number of channels increases, the stability is unchanged, and the enhancement in the peak ENOB is not obtained.

Table. ?? compares the two architectures with respect to the number of comparators and opamps needed for circuit implementation. The comparison was carried out by assuming the same bit resolutions of 18 bits and 23 bits for the first-order and

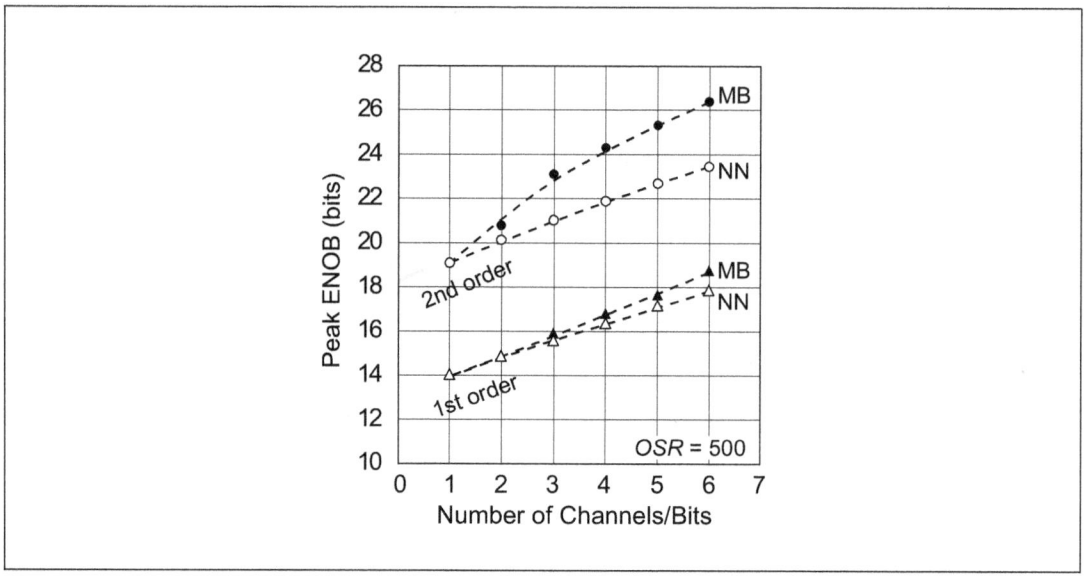

Figure 15: Peak ENOB as a function of the number of channels in the present ADCs using the first-order (△) and second-order (○) ΔΣ modulators (NN). For comparison, peak ENOB as a function of the number of bits in conventional first-order (▲) and second-order (●) multibit ΔΣ modulators are also plotted (MB).

	1st order @18 bits		2nd order @23 bits	
	Comparator	Opamp	Comparator	Opamp
This work	6	12	5 ~6	15 ~18
Multibit ΔΣ	32 (= 2^5)	1	8 (= 2^3)	2

Table 1: Comparison between this work and multibit ΔΣ modulator

second-order cases, respectively, based on the results shown in Fig. 15. In this work, to obtain a peak ENOB of 18 bits with the first-order configuration, 6 channels are required, which consists of 6 comparators, 6 opamps in the integrators, 5 opamps for the moving average, and 1 opamp in the summing node. A conventional 5-bit ΔΣ modulator was assumed for comparison. In this case, 2^5 comparators are necessary to obtain 5-bit resolution, but only one integrator is enough due to the single first-order negative feedback loop. As for the second order case, the number of channels is assumed to be 5 or 6 for this work, while 3-bit is assumed for a conventional multibit modulator. Therefore, the advantage of this work is the reduced number of comparators, while the drawback is the large number of opamps required here.

In addition, it should be noted that dynamic element matching (DEM) circuits are required to improve the linearity [15], because a digital-to-analog converter is used in the $\Delta\Sigma$ feedback loop. Such DEM circuits are not necessary in the present ADC, because it has only feedforward paths.

Since opamps consume much power, the number of opamps should be reduced by introducing novel circuit design techniques, such as ring amplifiers [16], dynamic-common-source circuits [17], and passive-mode circuits [18] [19]. These techniques are worth introducing into future circuit implementations. Another challenge might be designing the moving average circuit. The FIR filter based on the switched-capacitor technique [20] can be applied here because the moving average is a special case of the FIR filter, where the weighting coefficients are all the same, typically 1.

5 Conclusion

A novel analog-to-digital converter (ADC) architecture, which is based on a multi-channel feedforward network including $\Delta\Sigma$ modulators, was investigated. Multilevel signals were obtained by taking the moving average of the $\Delta\Sigma$ modulator output and used as feedforward signals. Our signal-level simulation revealed that the effective number of bits (ENOBs) increased by almost one bit at every one additional channel. An ENOB as high as 20 bits was predicted for the present 6-channel ADC with an oversampling ratio (OSR) of 250. This architecture is expected to provide a promising alternative approach for improving the ADC performance with a compact circuit configuration.

References

[1] M. Pelgrom, *Analog-to-Digital Conversion, 3rd Ed.* Springer, 2017.

[2] T. C. Carusone, D. Johns, and K. Martin, *Analog Integrated Circuit Design, 2nd Ed.* John Wiley & Sons, 2013.

[3] W. Kester, "A brief history of data conversion: A tale of nozzles, relays, tubes, transistors, and CMOS," *IEEE Solid-State Circuits Magazine*, vol. 7, no. 3, pp. 16–37, Summer 2015.

[4] D. Tank and J. Hopfield, "Simple 'neural' optimization networks: An A/D converter, signal decision circuit, and a linear programming circuit," *IEEE Trans. Circuits and Systems*, vol. 33, no. 5, pp. 533–541, May 1986.

[5] H. S. Lee, "Development of self-calibrating A/D converters," *IEEE Solid-State Circuits Magazine*, vol. 6, no. 2, pp. 18–21, Spring 2014.

[6] B. W. Lee and B. J. Sheu, "Design of a neural-based A/D converter using modified Hopfield network," *IEEE Journal of Solid-State Circuits*, vol. 24, no. 4, pp. 1129–1135, Aug 1989.

[7] T. Waho, "Analog-to-digital converters using not multi-level but multi-bit feedback paths," in *2017 IEEE 47th International Symposium on Multiple-Valued Logic (ISMVL)*, May 2017, pp. 7–12.

[8] ——, "An analog-to-digital converter using delta-sigma modulator network," in *2018 IEEE 48th International Symposium on Multiple-Valued Logic (ISMVL)*, May 2018, pp. 25–30.

[9] Y. Chigusa and M. Tanaka, "A neural-like feed-forward ADC," in *Circuits and Systems, 1990., IEEE International Symposium on*, May 1990, pp. 2959–2962 vol.4.

[10] G. Avitabile, M. Forti, S. Manetti, and M. Marini, "On a class of nonsymmetrical neural networks with application to ADC," *Circuits and Systems, IEEE Transactions on*, vol. 38, no. 2, pp. 202–209, Feb 1991.

[11] O. I. Zheng Tang and H. Matsumoto, "Implementing neural architectures using CMOS current-mode VLSI circuits," *IEICE Trans. Information and Systems*, vol. E74-D, no. 5, pp. 1329–1336, May 1991.

[12] R. Schreier and G. C. Temes, *Understanding Delta-Sigma Data Converters*. John Wiley & Sons, 2005.

[13] T. Waho, "Non-binary successive approximation analog-to-digital converters: A survey," in *2014 IEEE 44th International Symposium on Multiple-Valued Logic*, May 2014, pp. 73–78.

[14] T. Hayashi, Y. Inabe, K. Uchimura, and T. Kimura, "A multistage delta-sigma modulator without double integration loop," in *1986 IEEE International Solid-State Circuits Conference. Digest of Technical Papers*, vol. XXIX, Feb 1986, pp. 182–183.

[15] R. J. V. D. Plassche, "Dynamic element matching for high-accuracy monolithic D/A converters," *IEEE Journal of Solid-State Circuits*, vol. 11, no. 6, pp. 795–800, Dec 1976.

[16] B. Hershberg, S. Weaver, K. Sobue, S. Takeuchi, K. Hamashita, and U. K. Moon, "Ring amplifiers for switched capacitor circuits," *IEEE Journal of Solid-State Circuits*, vol. 47, no. 12, pp. 2928–2942, Dec 2012.

[17] R. Matsushiba, H. Kotani, and T. Waho, "An energy-efficient $\Delta\Sigma$ modulator using dynamic-common-source integrators," *IEICE Trans. Electronics*, vol. E97-C, no. 5, pp. 438–443, May 2014.

[18] W. Guo, H. Zhuang, and N. Sun, "A 13b-ENOB 173dB-FoM 2nd-order NS SAR ADC with passive integrators," in *2017 Symposium on VLSI Circuits*, June 2017, pp. C236–C237.

[19] C.-Y. Lin and T.-C. Lee, "A 12-bit 210-MS/s 5.3-mW pipelined-SAR ADC with a passive residue transfer technique," in *2014 Symposium on VLSI Circuits Digest of Technical Papers*, June 2014, pp. 1–2.

[20] G. Fischer, "Analog FIR filters by switched-capacitor techniques," *IEEE Transactions on Circuits and Systems*, vol. 37, no. 6, pp. 808–814, Jun 1990.

An Exact Optimization Method Using ZDDs for Linear Decomposition of Symmetric Index Generation Functions

Shinobu Nagayama
Dept. of Computer and Network Eng., Hiroshima City University, Hiroshima, JAPAN

Tsutomu Sasao
Dept. of Computer Science, Meiji University, Kawasaki, JAPAN

Jon T. Butler
Dept. of Electr. and Comp. Eng., Naval Postgraduate School, Monterey, CA USA

Abstract

This paper proposes a method using zero-suppressed binary decision diagrams (ZDDs) to find an exact optimum linear decomposition of symmetric index generation functions. The proposed optimization method recursively divides an index set of a symmetric index generation function, based on a branch and bound approach. The method uses ZDDs to represent partitions of an index set compactly and uniquely, and thus, it reuses partial solutions (partitions of an index set) efficiently to prune redundant solution search. In addition, by taking advantages of the symmetry property, the method reduces search space significantly, and can find an optimum solution quickly. Experimental results using benchmark symmetric index generation functions show effectiveness of the proposed method.

1 Introduction

Many network applications, such as detection of computer viruses and packet classification, use index searches as a basic operation. As network communication speeds increase, index searches have become a bottleneck. Especially, now virus patterns and classification rules need frequent updating. Thus, fast programmable *hardware* is essential in performing these index searches.

This paper is an extension of [5].

Index searches can be implemented as *Index generation functions* [7, 8]. To overcome the above challenges, an efficient memory-based hardware design method for index generation functions has been proposed [10]. The design method is based on *linear decomposition* [1, 6] of index generation functions, and decomposes an index generation function $f(x_1, x_2, \ldots, x_n)$ into two parts: L and G, as shown in Fig. 1. The first part L realizes linear functions y_i ($i = 1, 2, \ldots, p$) in a linear decomposition of f. L is realized by a programmable architecture [10] with EXOR gates, registers, and multiplexers. The second part G realizes a remaining function (general function) in a linear decomposition of f. G is realized by a $(2^p \times q)$-bit memory, where p is the number of linear functions, and q is the number of bits needed to represent function values of f.

In this design method, minimization of the number of linear functions, p, is indispensable to obtain a practical implementation because memory size of G strongly depends on p. Various minimization methods have been proposed [3, 4, 9, 10, 12, 13, 14, 15]. Most of them are heuristic, and as far as we know, only a few exact minimization methods [4, 5, 14, 15] have been proposed. Although heuristic methods are more scalable, devising an efficient exact minimization method is not only academically but also practically significant. This is because it becomes a basis for evaluating the quality of heuristic methods.

Since the exact minimization methods proposed in [14, 15] reduce the linear decomposition problem to a SAT problem, and solve it using a SAT solver, they no longer have much room for improvement unless the SAT solver is improved. On the other hand, the methods proposed in [4, 5] are emerging methods dedicated to solving the linear decomposition problem, and thus, they still have enough room for improvement. This paper focuses on improvement of the emerging methods, and proposes an exact optimization method for linear decomposition of symmetric index generation functions.

The proposed method uses zero-suppressed binary decision diagrams (ZDDs) to represent partial solutions of the problem compactly and uniquely. By using ZDDs, the proposed method can reuse partial solutions efficiently to prune redundant solution search. In addition, by taking advantages of the symmetry property of functions, the method reduces search space significantly, and can find an optimum solution much faster than the existing

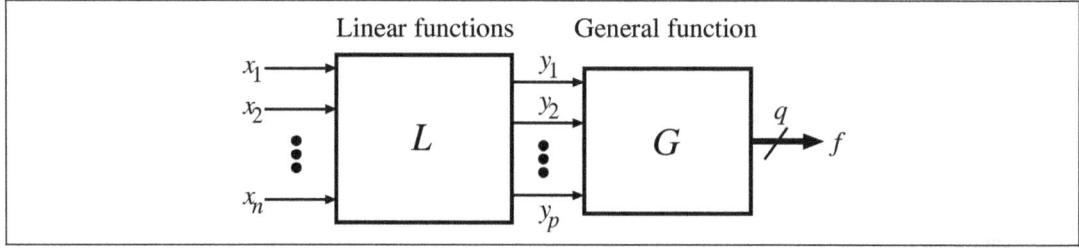

Figure 1: Linear decomposition of an index generation function [10].

methods [4, 5].

The rest of this paper is organized as follows: Section 2 defines symmetric index generation functions and linear decomposition. Section 3 formulates the optimization problem of linear decomposition, provides a brief overview of the existing exact optimization methods [4, 5], and proposes its improvement method using ZDDs and the symmetry property of functions. Section 4 shows experimental results using some benchmark symmetric index generation functions, and Section 5 concludes the paper.

2 Preliminaries

We briefly define index generation functions [7,8] and their linear decompositions [1,6,10].

Definition 1. *An **incompletely specified index generation function**, or simply **index generation function**, $f(x_1, x_2, \ldots, x_n)$ is a multi-valued function, where k assignments of values to binary variables $x_1, x_2, \ldots, x_n$ map to $K = \{1, 2, \ldots, k\}$. That is, the variables of f are binary-valued, while f is k-valued. Further, there is a one-to-one relationship between the k assignments of values to $x_1, x_2, \ldots, x_n$ and K. Other assignments are left unspecified. The k assignments of values to $x_1, x_2, \ldots, x_n$ are called the set of **registered vectors**. K is called the set of **indices**. $k = |K|$ is called the **weight** of the index generation function f.*

Example 1. *Fig. 2 shows 4-variable index generation functions with weight four. Note that, in Fig. 2(a), input values other than 0001, 0010, 0100, and 1101 are NOT assigned to any function values.* □

Definition 2. *A **characteristic function** χ of an index generation function $f(x_1, x_2, \ldots, x_n)$ is a logic function: $\{0,1\}^n \to \{0,1\}$ defined as*

$$\chi(x_1, x_2, \ldots, x_n) = \begin{cases} 1 & (f(x_1, x_2, \ldots, x_n) \in K) \\ 0 & (Otherwise). \end{cases}$$

Registered vectors				indices	Registered vectors				indices
x_1	x_2	x_3	x_4	f_a	x_1	x_2	x_3	x_4	f_s
0	0	0	1	1	1	0	0	0	1
0	0	1	0	2	0	1	0	0	2
0	1	0	0	3	0	0	1	0	3
1	1	0	1	4	0	0	0	1	4
(a) Asymmetric function.					(b) Symmetric function.				

Figure 2: Example of index generation functions [4].

Definition 3. *A **symmetric logic function** χ satisfies*

$$\chi(x_1,x_2,\ldots x_i,\ldots,x_j,\ldots,x_n) = \chi(x_1,x_2,\ldots x_j,\ldots,x_i,\ldots,x_n) \qquad (\forall x_i, x_j).$$

*In this function, function values are decided only by the number of 1's in an assignment of values to $x_1,x_2,\ldots,x_n$. An **elementary symmetric function** S_m^n is a special case of symmetric logic functions where $S_m^n = 1$ if and only if the number of 1's in an assignment to $x_1,x_2,\ldots,x_n$ is m.*

Definition 4. *Let $\chi(x_1,x_2,\ldots,x_n)$ be a characteristic function of an index generation function f. When χ is symmetric, f is a **symmetric index generation function**.*

Note that "f is a symmetric index generation function" does not mean f is a symmetric function (where any permutation of the input values leaves the function value unchanged). This is because, for each $e \in K$, there is only one assignment of values to the variables which maps to e. Therefore, only very few symmetric index generation functions are symmetric. An example of a symmetric index generation function that is symmetric is f_1, where $f_1(0,0,\ldots,0) = 1$, $f_1(1,1,\ldots,1) = 2$, and $f_1 = 0$ otherwise.

This paper considers only the case where χ is an elementary symmetric function. Thus, in this paper, *symmetric index generation function* means that its χ is an elementary symmetric function, unless otherwise stated.

Example 2. *Fig. 2(b) shows a symmetric index generation function.* □

Definition 5. *Let $K = \{1,2,\ldots,k\}$ be a set of indices of an index generation function. If $K = S_1 \cup S_2 \cup \ldots \cup S_u$, each $S_i \neq \emptyset$, and $S_i \cap S_j = \emptyset$ $(i \neq j)$, then $\mathcal{P} = \{S_1, S_2, \ldots, S_u\}$ is a **partition** of the set of indices K. When all the subsets S_i are singletons (i.e., $|S_i| = 1$), $|\mathcal{P}| = |K| = k$.*

An arbitrary n-variable index generation function with weight k can be realized by a $(2^n \times q)$-bit memory, where $q = \lceil \log_2(k+1) \rceil$. Linear decomposition is effective in reducing the memory size [10].

Definition 6. ***Linear decomposition** of an index generation function $f(x_1,x_2,\ldots,x_n)$ is a representation of f using a general function $g(y_1,y_2,\ldots,y_p)$ and linear functions y_i:*

$$y_i(x_1,x_2,\ldots,x_n) = a_{i1}x_1 \oplus a_{i2}x_2 \oplus \ldots \oplus a_{in}x_n$$
$$(i = 1,2,\ldots,p),$$

where $a_{ij} \in \{0,1\}$ $(j = 1,2,\ldots,n)$, and, for all registered vectors of the index generation function, the following holds:

$$f(x_1,x_2,\ldots,x_n) = g(y_1,y_2,\ldots,y_p).$$

y_1	y_2	g_1	g_2
0	0	1	2
0	1	2	1
1	0	3	3
1	1	4	4

Table 1: General functions g_1 and g_2 in linear decomposition of f_a [4].

Each y_i is called a **compound variable**. For each y_i, $\sum_{j=1}^{n} a_{ij}$ is called a **compound degree** of y_i, denoted by $deg(y_i)$, where a_{ij} is viewed as an integer, and $\sum$ as an integer sum.

Definition 7. *An inverse function of a general function $z = g(y_1, y_2, \ldots, y_p)$ in a linear decomposition is a mapping from $K = \{1, 2, \ldots, k\}$ to a set of p-bit vectors $\{0,1\}^p$, denoted by $g^{-1}(z)$. In this inverse function $g^{-1}(z)$, a mapping obtained by focusing only on the i-th bit of the p-bit vectors: $K \to \{0,1\}$ is called an **inverse function to a compound variable** y_i, denoted by $(g^{-1})_i(z)$.*

Definition 8. *Let $ON(y_i) = \{z \mid z \in K, (g^{-1})_i(z) = 1\}$, where $K = \{1, 2, \ldots, k\}$ and $(g^{-1})_i(z)$ is an inverse function of $g(y_1, y_2, \ldots, y_n)$ to y_i. $|ON(y_i)|$ is called the **cardinality of** y_i or informally the **number of 1s included in** y_i.*

Example 3. *The index generation function f_a in Example 2 can be represented by $y_1 = x_2$, $y_2 = x_1 \oplus x_3$, and $g_1(y_1, y_2)$ shown in Table 1. In this case, $deg(y_1) = 1$ and $deg(y_2) = 2$, respectively. f_a can be also represented by $y_1 = x_2$, $y_2 = x_4$, and $g_2(y_1, y_2)$ in the same table. In this case, both $deg(y_1)$ and $deg(y_2)$ are 1. In either case, f_a can be realized by the architecture in Fig. 1 with a $(2^2 \times 3)$-bit memory.*

For $g_2(y_1, y_2)$ in Table 1, its inverse functions to y_1 and y_2 are $(g_2^{-1})_1(z)$ and $(g_2^{-1})_2(z)$, respectively. We have $(g_2^{-1})_1(2) = 0$, $(g_2^{-1})_1(1) = 0$, $(g_2^{-1})_1(3) = 1$, and $(g_2^{-1})_1(4) = 1$. Similarly, $(g_2^{-1})_2(2) = 0$, $(g_2^{-1})_2(1) = 1$, $(g_2^{-1})_2(3) = 0$, and $(g_2^{-1})_2(4) = 1$. The cardinalities of both y_1 and y_2 are 2. □

In this way, linear decomposition can significantly reduce memory size needed to realize an index generation function. But, in linear decomposition, not only memory, but also EXOR gates, registers, and multiplexers are required to realize a compound variable (i.e., block L in Fig. 1). Since circuit size of L depends on compound degrees, lower compound degrees are desirable when the memory size is equal.

3 Exact Optimization of Linear Decomposition

This section formulates the optimization problem of linear decomposition, and shows exact optimization methods to solve the problem.

3.1 Formulation of Optimization Problem

Linear decomposition of an index generation function is realized by the architecture in Fig. 1 using EXOR gates, registers, multiplexers, and a $(2^p \times q)$-bit memory. Among these components, only memory requires size exponentially growing with the number of linear functions p. To reduce memory size, we address the following problem:

Problem 1. *Given an index generation function f and an integer t, find a linear decomposition of f such that the number of linear functions p is minimum, and compound degrees are at most t.*

The constraint on compound degrees t is given to constrain not only solution space, but also delay and area of the circuit L realizing the linear functions.

Example 4. *For linear decompositions of f_a in Example 3, the decomposition with $y_1 = x_2$, $y_2 = x_4$, and $g_2(y_1, y_2)$ is optimum when $t = 1$.* □

3.2 Existing Methods Based on Partition of Indices

This subsection shows a brief review of existing exact optimization methods [4, 5] that are basis of the proposed method.

3.2.1 Overview of Existing Methods

We consider Problem 1 as a problem of minimizing the height of a binary decision tree constructed by compound variables [3].

Example 5. *Fig. 3 shows a binary decision tree of the smallest height that divides the set of indices into singletons by compound variables y_1 and y_2. It corresponds to g_2 in Table 1.* □

The existing methods [4, 5] search for a binary decision tree with the smallest height, based on a branch and bound approach. They select a candidate of an optimum compound variable one by one in a top-down manner, and divide a given set of indices recursively by selected variables while constructing a binary decision tree. By comparing heights of the trees, the best combination of compound variables is found. Algorithm 1 shows the overview of the branch and bound approach.

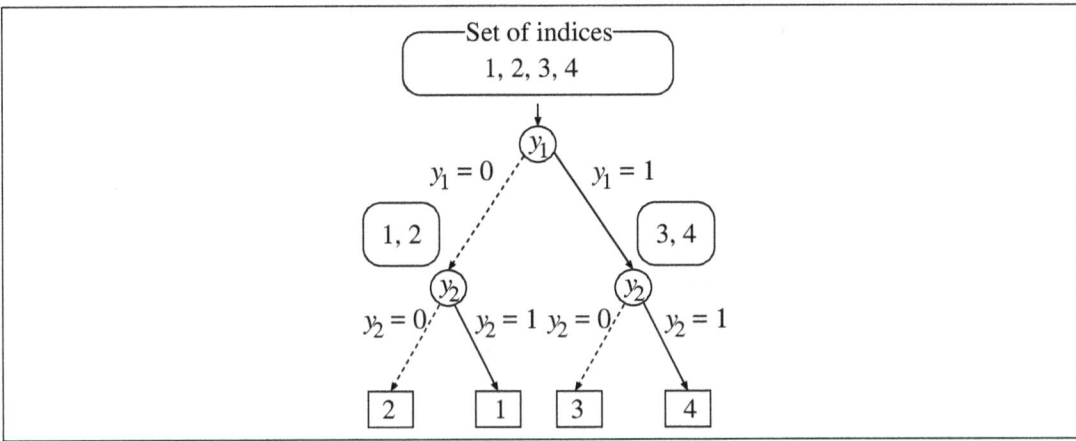

Figure 3: Binary decision tree for g_2 of Table 1 [5].

Algorithm 1. *Overview of the branch and bound approach in [4, 5]*

Input: an index generation function with weight k and a compound degree t
Output: a set of compound variables and its size h_{min}

Let $\mathcal{P} = \{K\}, h = 0$, and iterate the following recursively.
min_search($\mathcal{P}, h$) {
   if ($|\mathcal{P}| = k$) { **update_solution**(h); return; }
   if (**bound_condition**($\mathcal{P}, h$) is satisfied) return;
   branch($\mathcal{P}, t, h$);
}

Algorithm 1 searches for a solution recursively while constructing a binary decision tree with height h. When $|\mathcal{P}| = k$ (i.e., the set of indices is divided into singletons), a solution (a set of h compound variables) is obtained. The procedure *update_solution*() compares the obtained solution with the current solution, and updates the current solution if the obtained one is better. The procedure *branch*() explores the solution space by selecting a compound variable using two cost functions proposed for the heuristic method [3]:

$$cost_1(\mathcal{P}, y_i) = \sqrt{\sum_{S \in \mathcal{P}} \left(\frac{|S|}{2} - |S \cap ON(y_i)| \right)^2}$$

and

$$cost_2(\mathcal{P}, y_i) = \max_{S \in \mathcal{P}} \{\max\{|S \cap ON(y_i)|, |S \setminus ON(y_i)|\}\},$$

where $\mathcal{P}$ is a partition of a set of indices with already selected compound variables. The procedure *bound_condition()* detects an ineffective solution using the lower bound discussed below in Theorem 1, and prunes it.

Theorem 1. *[4] Let m be the number of indices in a set, and c be the number of 1s in compound variables. When $c < \frac{m}{2}$, at least*

$$lower(m,c) = \left\lfloor \frac{m}{c} \right\rfloor + \lceil \log_2(c) \rceil - 1$$

compound variables are needed to divide the set into m singletons.

3.2.2 Previous Improvement Method Using ZDDs [5]

Before describing the improvement method using ZDDs [5], we briefly define ZDDs.

Definition 9. *A zero-suppressed binary decision diagram (ZDD) [2] is a rooted directed acyclic graph (DAG) representing a logic function. It consists of two terminal nodes representing function values 0 and 1, and nonterminal nodes representing input variables. Each nonterminal node has two outgoing edges, 0-edge and 1-edge, that correspond to the values of the input variables. Neither terminal node has outgoing edges.*

A ZDD is obtained by repeatedly applying the Shannon expansion $f = \overline{x_i} f_0 \vee x_i f_1$ to a logic function, where $f_0 = f(0 \to x_i)$, and $f_1 = f(1 \to x_i)$, and by applying the following two reduction rules:

1. *Coalesce equivalent sub-graphs.*

2. *Delete nonterminal nodes whose 1-edge points to the terminal node representing 0, and redirect edges that point to the deleted node, to the node, to which the 0-edge of the deleted node has pointed.*

As is well known, ZDDs represent partitions compactly and uniquely [2], and thus, a partition of an index set $\mathcal{P} = \{S_1, S_2, \ldots, S_u\}$ can be also represented compactly and uniquely using a ZDD. Example 6 below shows an example of the use of ZDDs.

Example 6. *Let an index set be $K = \{1,2,3,4,5,6\}$, and a partition of K be $\mathcal{P} = \{\{1,3,6\}, \{2,5\}, \{4\}\}$. Fig. 4 shows a ZDD for $\mathcal{P}$. In Fig. 4, dashed lines and solid lines denote 0-edges and 1-edges, respectively. The number of nonterminal nodes is 6.* □

Theorem 2. *[5] Let an index set be $K = \{1,2,\ldots,k\}$. For any partition $\mathcal{P}$ of K, the number of nonterminal nodes in a ZDD for $\mathcal{P}$ is k, regardless of the variable order.*

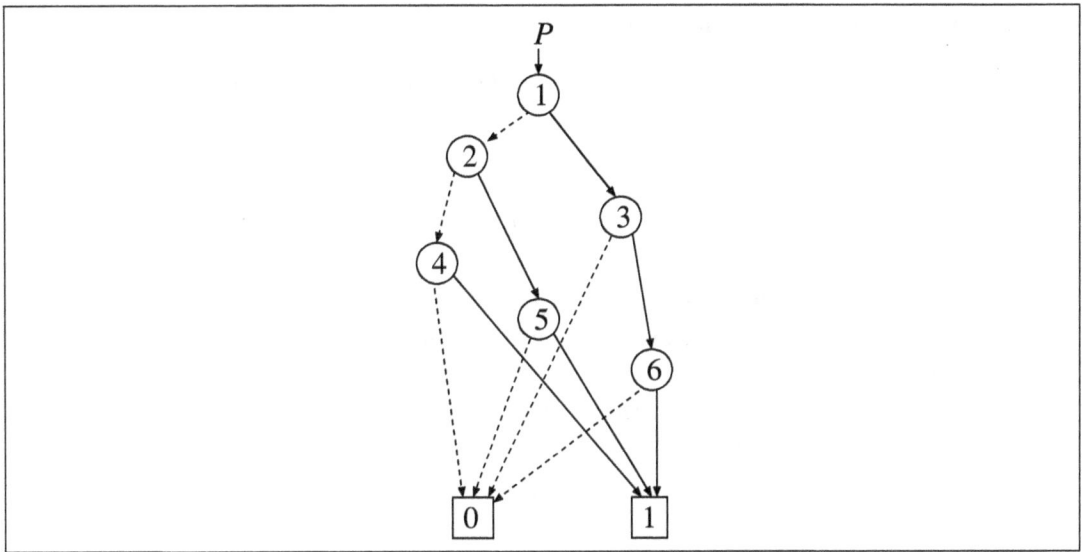

Figure 4: ZDD for $\mathcal{P} = \{\{1,3,6\},\{2,5\},\{4\}\}$ [5].

As described in Section 3.2.1, Algorithm 1 searches for an optimum solution while dividing a given index set repeatedly. Thus, a same partition of indices tends to repeatedly appear during solution search. However, once the minimum number of compound variables needed to divide a partition of indices into singletons is found, we do not need to divide the same partition again to obtain its minimum number of compound variables. By reusing subsolutions obtained by past solution search, we can prune such redundant search space. The question is how to store subsolutions (i.e. partitions of indices) compactly and uniquely in memory. To answer this question, we use ZDDs.

Algorithm 2 shows the overview of the improved method using ZDDs. A ZDD is constructed for each partition using *Change* and *Union* operations that are basic operations in a ZDD package [2]. Then, the ZDD is checked to determine whether the current subsolution has already been obtained. This checking is made by searching the history of ZDDs for the equivalent one. Such a search (equivalence checking) is what ZDDs do best; it is done in $O(1)$ time. If the subsolution has already been obtained, and it does not improve the current solution, then the solution search is pruned. On the other hand, when any subsolution has not been obtained yet, the solution search is performed, and after that, the ZDD is stored to the history with the obtained subsolution. In this way, redundant solution search is efficiently pruned using ZDDs.

Algorithm 2. *Overview of the improved method using ZDDs*

> Input: an index generation function with weight k and a compound degree t
> Output: a set of compound variables and its size h_{min}

Let $\mathcal{P} = \{K\}, h = 0$, and iterate the following recursively.
min_search($\mathcal{P}, h$) {
  if ($|\mathcal{P}| = k$) { update_solution(h); return; }
  Construct a ZDD for $\mathcal{P}$;
  Search for ZDD_subsolution in ZDD_History();
  if ((ZDD_subsolution has been obtained) and (its min_lower + $h \geq h_{min}$))
    return;
  if (bound_condition($\mathcal{P}, h$) is satisfied) return;
  min_lower = branch($\mathcal{P}, t, h$);
  Add the ZDD and min_lower to ZDD_History;
}

3.3 Proposed Improvement Method Using ZDDs and Symmetry Property

The method described in Section 3.2.2 can be improved further by targeting symmetric index generation functions. This subsection proposes an improvement method using ZDDs in combination with the symmetry property of functions.

In Section 3.2.2, ZDDs *directly* represent partitions of indices, and the ZDDs are used to prune redundant solution search. However, for symmetric index generation functions, each value of index itself has no meaning, and thus, it can be abstracted. Each index can be used only to distinguish from others. This is because indices of symmetric index generation functions are exchangeable when Problem 1 is considered. Thus, in a partition of indices, we define a *relation* between size of subset in the partition and the number of subsets with the same size, and then, we represent the relation using a ZDD.

Example 7. *In a partition of indices* $\{\{1,3,6\}, \{2,5\}, \{4\}\}$, *size of each subset is 3, 2, and 1. Thus, we have a relation* $\{(3,1),(2,1),(1,1)\}$. *In another partition of indices* $\{\{1,2\}, \{3,4\},\{5,6\}\}$, *size of all the three subsets is 2. Thus, we have a relation* $\{(2,3)\}$. □

In this way, for symmetric index generation functions, we can remove index values when considering the minimum number of compound variables needed to divide a partition of indices into singletons. This is because only sizes of subsets are important. By using this representation to remove index values, partitions of indices can be classified into *equivalence classes*. Since a ZDD represents a *representative* of each equivalence class, we can prune redundant solution search significantly.

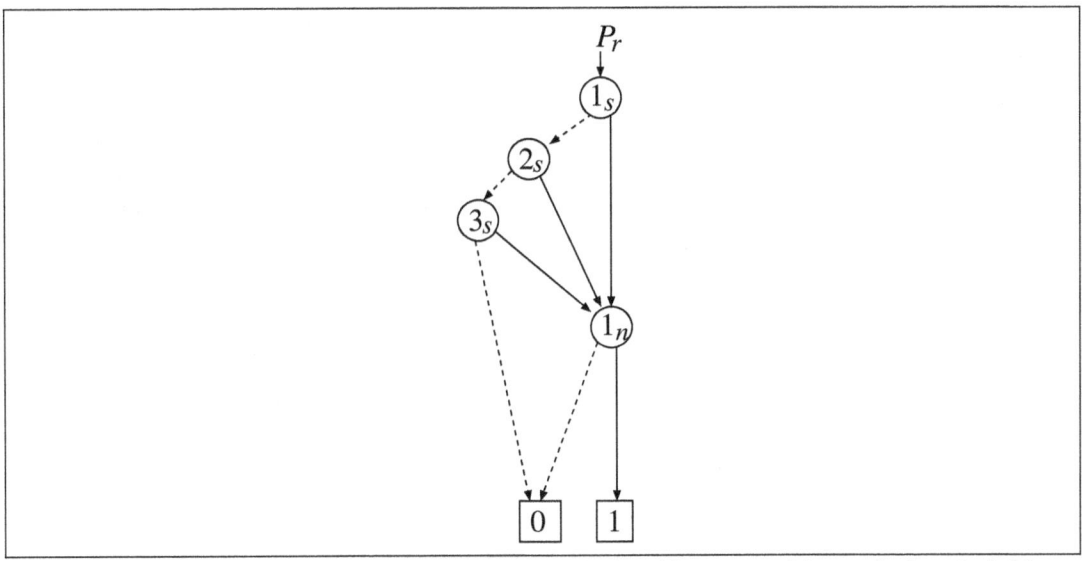

Figure 5: ZDD for a relation $\mathcal{P}_r = \{(3,1),(2,1),(1,1)\}$ of $\mathcal{P} = \{\{1,3,6\},\{2,5\},\{4\}\}$.

Example 8. *Fig. 5 shows a ZDD for a relation $\mathcal{P}_r = \{(3,1),(2,1),(1,1)\}$ of a partition of indices $\mathcal{P} = \{\{1,3,6\}, \{2,5\},\{4\}\}$ in Example 7. In Fig. 5, 1_s, 2_s, and 3_s denote subset sizes 1, 2, and 3, respectively. And, 1_n denotes that the number of subsets with the same size is 1. Since the size of each subset is different than others, all the pairs in $\mathcal{P}_r$ share the nonterminal node of 1_n in the ZDD.* □

Theorem 3. *Let an index set be $K = \{1,2,\ldots,k\}$, a partition of K be $\mathcal{P}$, and a relation between subset size and the number of subsets in $\mathcal{P}$ be $\mathcal{P}_r$. An upper bound on the number of nonterminal nodes in a ZDD for $\mathcal{P}_r$ is $\sqrt{8k+1}-1$.*

Proof: Let $\mathcal{P} = \{S_1, S_2, \ldots, S_u\}$. The number of elements in $\mathcal{P}_r$ is maximum when the size of each subset S_i differs from each other. Without loss of generality, let $|S_1| = 1, |S_2| = 2, \ldots, |S_u| = u$. Then, the total number of elements in all S_i's is

$$\sum_{i=1}^{u} i = \frac{u(u+1)}{2}.$$

Since $\mathcal{P}$ is a partition of indices, the total number of elements is not larger than k, and we have

$$\frac{u(u+1)}{2} \leq k.$$

From $u \geq 0$ and $k \geq 0$, we have

$$u \leq \frac{\sqrt{8k+1}-1}{2}. \tag{1}$$

Let a **1-path** in a ZDD be a sequence of edges and nodes leading from the root node to the terminal node representing 1. A 1-path in a ZDD for $\mathcal{P}_r$ represents an ordered pair in $\mathcal{P}_r$, and a pair of a 1-edge and its nonterminal node on the 1-path represents an element in the ordered pair. Thus, the number of 1-paths is exactly $|\mathcal{P}_r|$, and the number of pairs of a 1-edge and its nonterminal node on each 1-path is exactly 2. Unless nonterminal nodes are shared, the number of nonterminal nodes is $|\mathcal{P}_r| \times 2$ From (1), we have the theorem. □

While the number of nonterminal nodes in a ZDD directly representing a partition of indices $\mathcal{P}$ is $O(k)$, the number of nodes in a ZDD representing a relation of $\mathcal{P}$ is $O(\sqrt{k})$. Thus, this method improves not only search time but also memory size.

4 Experimental Results

The proposed exact minimization method is implemented in the C language, and run on the following computer environment: CPU: Intel Core2 Quad Q6600 2.4GHz, memory: 4GB, OS: CentOS 5.7 Linux, and C-compiler: gcc -O2 (version 4.1.2).

4.1 On Reduction of Search Space

To evaluate the effectiveness of the proposed improvement method, we compare the proposed method with the existing methods [4, 5] in terms of search space size. Table 2 shows the number of times that the procedure *branch()* is invoked in each method for some benchmark symmetric index generation functions shown in [10]. The bold values in Table 2 show where the proposed method significantly outperforms the other two methods.

As shown in Table 2, the search space size of the proposed method using ZDDs in combination with the symmetry property of functions is several orders of magnitude smaller than the search space of the existing ones. Search space is reduced by using ZDDs, as achieved by the existing method of [5]. And, by taking advantages of the symmetry property together, effectiveness of ZDDs is significantly improved. Even for functions where the existing methods could not find the optimum solution because of large search space, the proposed method can find the optimum solution in a short computation time by avoiding redundant solution searching. From these results, we can see that the proposed improvement method has a significant effect on the reduction of the redundant solution searching.

4.2 On Number of Nonterminal Nodes

Table 3 shows the total number of nonterminal nodes in ZDDs needed to represent all partitions of an index set that appeared during a search for solutions in the proposed method and the existing method [5]. Note that this number does not include unused nodes after ZDD

OPTIMIZATION FOR LINEAR DECOMPOSITION OF IGFS

Benchmark functions	Compound degrees	h_{min}	Number of times *branch()* is invoked		
			Existing method [4]	Existing method [5]	Proposed method
1-out-of-10	$t=1$	9	9	9	9
	$t=2$	6	1,975,364	5,310	**13**
	$t=3$	5	151,773	2,268	**6**
	$t=4$	4	4	4	4
	$t=5$	4	4	4	4
1-out-of-12	$t=1$	11	11	11	11
	$t=2$	8	†	†	**29**
	$t=3$	6	†	†	**16**
	$t=4$	5	†	6,274	**6**
	$t=5$	4	†	†	**4**
1-out-of-16	$t=1$	15	15	15	15
	$t=2$	11	†	†	**57**
	$t=3$	8	†	†	**78**
	$t=4$	6	†	†	**10**
	$t=5$	5	5	5	5
2-out-of-16	$t=1$	15	†	†	**15**
	$t=2$	11	†	†	**62**
	$t=3$	9	†	†	**26**
	$t=4$	8	8	8	8
	$t=5$	8	9	9	9
3-out-of-16	$t=1$	15	†	†	**15**
	$t=2$	13	†	†	**142**
	$t=3$	11	†	†	**113**
	$t=4$	10	10	10	10
	$t=5$	10	†	†	**32,239**

† Computation was terminated when it exceeded one hour.

Table 2: Comparison of methods in terms of search space.

operations (that is, Table 3 shows the number of nodes after *garbage collection* is applied). These results correspond to space (memory size) complexities of both the methods.

Since a ZDD is constructed every time *branch()* is invoked, the number of ZDDs is equal to the search space shown in Table 2. As shown in Table 3, the number of ZDDs in the proposed method is smaller than that in the existing method because the search space of

Benchmark functions	Compound degrees	k	Existing method [5]		Proposed method	
			No. of ZDDs	No. of nodes	No. of ZDDs	No. of nodes
1-out-of-10	$t=1$	10	9	59	9	26
	$t=2$	10	5,310	11,564	13	33
	$t=3$	10	2,268	4,765	6	19
	$t=4$	10	4	36	4	14
	$t=5$	10	4	38	4	12
1-out-of-12	$t=1$	12	11	84	11	32
	$t=2$	12	†	†	29	72
	$t=3$	12	†	†	16	43
	$t=4$	12	6,274	13,451	6	19
	$t=5$	12	†	†	4	15
1-out-of-16	$t=1$	16	15	144	15	44
	$t=2$	16	†	†	57	136
	$t=3$	16	†	†	78	187
	$t=4$	16	†	†	10	27
	$t=5$	16	5	67	5	18
2-out-of-16	$t=1$	120	†	†	15	62
	$t=2$	120	†	†	62	262
	$t=3$	120	†	†	26	102
	$t=4$	120	8	939	8	44
	$t=5$	120	8	1,069	9	47
3-out-of-16	$t=1$	560	†	†	15	77
	$t=2$	560	†	†	142	910
	$t=3$	560	†	†	113	979
	$t=4$	560	10	5,595	10	74
	$t=5$	560	†	†	32,239	186,256

† Computation was terminated when it exceeded one hour.

Table 3: Total number of nonterminal nodes in ZDDs.

the proposed method is smaller. In addition, each ZDD in the proposed method has fewer nodes than one in the existing method. Therefore, not only theoretical space complexity shown in Theorem 3 but also practical space complexity of the proposed method is much lower than that of the existing method.

Benchmark functions	Compound degrees	Existing method [4]	Existing method [5]	Proposed method
1-out-of-10	$t=1$	*<0.01	*<0.01	*<0.01
	$t=2$	216.62	0.84	***<0.01**
	$t=3$	127.23	1.12	***<0.01**
	$t=4$	0.02	*<0.01	*<0.01
	$t=5$	0.14	*<0.01	*<0.01
1-out-of-12	$t=1$	*<0.01	*<0.01	*<0.01
	$t=2$	†	†	***<0.01**
	$t=3$	†	†	***<0.01**
	$t=4$	†	18.34	***<0.01**
	$t=5$	†	†	***<0.01**
1-out-of-16	$t=1$	*<0.01	*<0.01	*<0.01
	$t=2$	†	†	**0.01**
	$t=3$	†	†	**0.09**
	$t=4$	†	†	**0.04**
	$t=5$	3.68	0.18	**0.06**
2-out-of-16	$t=1$	†	†	***<0.01**
	$t=2$	†	†	**0.07**
	$t=3$	†	†	**0.17**
	$t=4$	2.76	5.24	**0.17**
	$t=5$	36.32	14.13	**0.52**
3-out-of-16	$t=1$	†	†	**0.01**
	$t=2$	†	†	**0.75**
	$t=3$	†	†	**2.96**
	$t=4$	16.17	207.65	**0.95**
	$t=5$	†	†	**479.05**

* Time is less than 0.01 sec..
† Computation was terminated when it exceeded one hour.

Table 4: Computation time of methods (in seconds).

4.3 On Computation Time

Although the proposed improvement method reduces search space significantly, as shown in Table 2, computational overhead can negate such an improvement. To show that the overhead is small enough and reduction of search space leads to shortening of computation time, we compare computation time of the three methods. Table 4 shows computation time,

Benchmark functions	Compound degrees	k	h_{min}	Space	No. of nodes	Time (sec.)
1-out-of-20	$t=1$	20	19	19	56	*<0.01
	$t=2$	20	13	91	203	0.03
	$t=3$	20	10	196	439	0.53
	$t=4$	20	8	149	348	1.92
	$t=5$	20	7	62	163	2.78
2-out-of-20	$t=1$	190	19	19	82	*<0.01
	$t=2$	190	14	189	784	0.50
	$t=3$	190	12	1,732	8,938	34.39
	$t=4$	190	10	78	440	5.73
	$t=5$	190	9	10	55	2.66
3-out-of-20	$t=1$	1,140	19	19	105	0.02
	$t=2$	1,140	16	414	2,866	6.93
	$t=3$	1,140	13	328	4,046	33.38
	$t=4$	1,140	12	1,866	11,122	1,137.29
4-out-of-20	$t=1$	4,845	19	19	118	0.12
	$t=2$	4,845	16	129	1,195	8.36

* Time is less than 0.01 sec..

Table 5: Experimental results for larger benchmark functions.

in seconds, of the three methods for the same benchmark functions.

For "2-out-of-16" with $t = 4$ and "3-out-of-16" with $t = 4$, the existing method using ZDDs [5] is slower than the method without using ZDDs [4]. This is because, for these functions, search space is not reduced at all, and *garbage collection* is applied frequently due to many unused nodes produced as a by-product of ZDD operations. However, the proposed method is faster than the existing method [4] even though ZDD operations are performed. This is because the number of nodes in the proposed method is smaller.

These results show that the proposed method using ZDDs in combination with the symmetry property of functions is effective in reducing overheads of ZDD operations as well as search space, resulting in fast computation.

4.4 For Larger Benchmark Functions

Since the proposed method is much faster than the existing methods, we applied it to larger benchmark symmetric index generation functions. Here, existing methods cannot find an optimum solution in a reasonable computation time due to a huge search space. Table 5 shows their results.

From these results, we can see that the proposed method is promising for the problem of finding an optimum solution of large symmetric index generation functions.

5 Conclusion and Comments

This paper proposes an exact optimization method for linear decomposition of symmetric index generation functions. By taking advantages of the symmetry property of functions in combination with ZDDs, space of solution search and computational overhead of ZDD operations are reduced significantly. Thus, the proposed method can find an optimum solution for symmetric index generation functions that the existing methods could not find. Experimental results show that *time and space complexities* of the proposed method for symmetric index generation functions are reasonable.

Since the proposed method quickly finds an exact optimum design for symmetric index generation functions, *circuit complexities* of L and G in linear decomposition for the functions can be analyzed more precisely. Investigating relations between those circuit complexities and the compound degree t would be interesting.

Acknowledgments

This research is partly supported by the JSPS KAKENHI Grant (C), No.16K00079, 2018. We would like to thank Prof. Michael Miller for motivating us to use the symmetry property. The reviewers' comments were helpful in improving the paper.

References

[1] R. J. Lechner, "Harmonic analysis of switching functions," in A. Mukhopadhyay (ed.), *Recent Developments in Switching Theory*, Academic Press, New York, Chapter V, pp. 121–228, 1971.

[2] S. Minato, "Zero-suppressed BDDs for set manipulation in combinatorial problems," *Proc. 30th Design Automation Conference*, pp. 272–277, 1993.

[3] S. Nagayama, T. Sasao, and J. T. Butler, "An efficient heuristic for linear decomposition of index generation functions," *46th International Symposium on Multiple-Valued Logic*, pp. 96–101, May, 2016.

[4] S. Nagayama, T. Sasao, and J. T. Butler, "An exact optimization algorithm for linear decomposition of index generation functions," *47th International Symposium on Multiple-Valued Logic*, pp. 161–166, May, 2017.

[5] S. Nagayama, T. Sasao, and J. T. Butler, "An exact optimization method using ZDDs for linear decomposition of index generation functions," *48th International Symposium on Multiple-Valued Logic*, pp. 144–149, May, 2018.

[6] E. I. Nechiporuk, "On the synthesis of networks using linear transformations of variables," *Dokl, AN SSSR*, Vol. 123, No. 4, pp. 610–612, Dec., 1958 (in Russian).

[7] T. Sasao, *Memory-Based Logic Synthesis*, Springer, 2011.

[8] T. Sasao, "Index generation functions: recent developments (invited paper)," *41st International Symposium on Multiple-Valued Logic*, pp. 1–9, May 2011.

[9] T. Sasao, "Linear transformations for variable reduction," *Reed-Muller Workshop 2011*, May 2011.

[10] T. Sasao, "Linear decomposition of index generation functions," *17th Asia and South Pacific Design Automation Conference*, pp. 781–788, Jan. 2012.

[11] T. Sasao, Y. Urano, and Y. Iguchi, "A lower bound on the number of variables to represent incompletely specified index generation functions," *44th International Symposium on Multiple-Valued Logic*, pp. 7–12, May 2014.

[12] T. Sasao, Y. Urano, and Y. Iguchi, "A method to find linear decompositions for incompletely specified index generation functions using difference matrix," *IEICE Transactions on Fundamentals*, Vol. E97-A, No. 12, pp. 2427–2433, Dec. 2014.

[13] T. Sasao, "A reduction method for the number of variables to represent index generation functions: s-min method," *45th International Symposium on Multiple-Valued Logic*, pp. 164–169, May 2015.

[14] T. Sasao, I. Fumishi, and Y. Iguchi, "A method to minimize variables for incompletely specified index generation functions using a SAT solver," *International Workshop on Logic and Synthesis*, pp. 161–167, June 2015.

[15] T. Sasao, I. Fumishi, and Y. Iguchi, "On an exact minimization of variables for incompletely specified index generation functions using SAT," *Note on Multiple-Valued Logic in Japan*, Vol.38, No.3, pp. 1–8, Sept. 2015 (in Japanese).

www.ingramcontent.com/pod-product-compliance
Lightning Source LLC
Chambersburg PA
CBHW080552170426
43195CB00016B/2770